K. V. Laurikainen · The Message of the Atoms
Essays on Wolfgang Pauli and the Unspeakable

Springer
Berlin
Heidelberg
New York
Barcelona
Budapest
Hong Kong
London
Milan
Paris
Santa Clara
Singapore
Tokyo

KALERVO V. LAURIKAINEN

The Message of the Atoms

Essays
on Wolfgang Pauli
and the
Unspeakable

 Springer

Prof. Kalervo V. Laurikainen
Kelotie 4 C
FIN-01820 Klaukkala, Finland

The first Finnish edition, "Atomien viesti",
was published by Yliopistopaino/Helsinki University Press in 1994

ISBN-13:978-3-642-64457-3 Springer-Verlag Berlin Heidelberg New York

Library of Congress Cataloging-in Publication Data.
Laurikainen, Kalervo Vihtori.
The message of the atoms : Wolfgang Pauli and the unspeakable / K. V. Laurikainen. p. cm.
ISBN-13:978-3-642-64457-3   e-ISBN-13:978-3-642-60560-4
DOI: 10.1007/978-3-642-60560-4

1. Quantum theory. 2. Physics – Philosophy. 3. Pauli, Wolfgang, 1900–1958. I. Title.
QC174.12.L39 1996 530.1'2–dc20 96-43199 CIP

© Springer-Verlag Berlin Heidelberg 1997
Softcover reprint of the hardcover 1st edition 1997

Cover design: design & production GmbH, Heidelberg
Typesetting: Data conversion by Text & Grafik GmbH, Heidelberg
SPIN 10543856   55/3142– 5 4 3 2 1 0 – Printed on acid-free paper

Prologue

This book is an attempt to interpret the message of the atoms as Wolfgang Pauli interpreted it, but perhaps in a way which is easier to understand. Errors are to be ascribed to the author, not to Pauli.

Man can guide another only to the shore of the Styx. Sometimes one can feel, however, that Pauli has found a scenic place from which something can be dimly seen even in the world beyond.

In the chapter "Numinosum" the story about what I see from this scenic place begins.

"*They went a little too far in the seventeenth century*"
(From a letter of Wolfgang Pauli to Markus Fierz, October 13, 1951.)

Preface

> *When I meet a difficult problem, I begin to go around it, approaching it again and again from different directions. If I persistently continue these approaches, it can happen that no problem remains.*
> (Rolf Nevanlinna, in a private discussion.)

In 1976, after a mainly administrative period of some 15 years, I spent a couple of months at CERN, working in the Pauli Collection. When I found the Pauli-Fierz correspondence, I had the intuitive feeling that there was the key: that „it was an objective description, and that it was the only possible objective description" for the mysteries of quantum mechanics. Here I have cited Bohr in his 'last interview' (see Chap. 7), which I became acquainted with only later, but I was immediately convinced that Pauli's view was more profound than anything else I had read about in quantum mechanics.

However, nowadays the investigation of the foundations of quantum theory is dominated by 'realism', which means that the influence of the psyche on our conception of reality is ignored. This book is an attempt to show that this is not possible in quantum mechanics.

Contrary to Bohr, Pauli did not avoid the discussion of ontological implications of quantum mechanics, and he found in C.G. Jung's *unus mundus* a psychological counterpart to his views. Pauli writes of a *cosmic order* which is beyond the distinction of the physical and the psychic aspects of reality – it means a fusion of the 'outer world' and the 'inner world'. Such considerations led Pauli to the borderland between knowledge and belief, and perhaps this explains the repression of his philosophical thought among colleagues in physics.

A new, *quaternarian*, view of science was characteristic of Pauli in his later years. This is in harmony with the holistic nature of quantum theory. The quaternarian attitude emphasizes interdisciplinary wholes, while the present, 'trinitarian', science results in more and more isolated disciplines. Important is a harmonious relation between science and religion, in the spirit of Pauli's remark in his letter of August 12, 1948, to Fierz:

Ich meine nicht „Religion innerhalb der Physik" und auch nicht „Physik innerhalb der Religion" – denn beides wäre ja „einseitig" – sondern Einordnung beider in ein Ganzes.

[I do not mean „religion within physics", nor do I mean „physics inside of religion", since either one would certainly be „one-sided", but rather I mean the placing of both of them within a whole.]

In the present science-religion dialogue, Pauli's view opens an important perspective which is essentially different from the 'realistic' approaches to quantum mechanics mentioned above.

I have the feeling that these trends in Pauli's thought correspond to the present needs of cultural development. Therefore, Pauli's philosophy should be better known, not only among physicists and philosophers but within a much wider audience. Perhaps the time is dawning when complementarity will be taught in elementary schools – as Bohr once predicted.

This book is an attempt to speak about the Unspeakable. Therefore it is not 'linear', proceeding consistently from a certain beginning to a certain end. Rather, it is a collection of approaches towards the Unvisible – in the hope that the reader will step by step begin to see a structure in reality which is beyond our reach. This structure gives direction to our endeavors.

Part II, „Facts and Interpretations", differs essentially in its style from the other three parts: it is more like a customary scientific treatise. It is, in fact, a translation of four chapters from another book, *Tieteellä on rajansa* (Science Has Its Limits), and three of these chapters were originally written for scientific journals. The aim of Part II is to show that the philosophy of the Copenhagen interpretation is generally misunderstood today. In Chap. 5, citations from original articles (partly in German) are used for documentation.

I hope to convince the reader of the incompatibility of quantum mechanics with 'realism' if this is understood in the usual physical sense, without incorporating consciousness into the conception of reality. If this incompatibility is not understood, the 'mysteries' of quantum mechanics cannot be eliminated – other than by theories that can be neither verified nor falsified. It is time to reject the 'realistic' misunderstandings and to listen again to Bohr and Pauli.

Acknowledgements

I wish to express my gratitude to a very active group of people in the Finnish Society for Natural Philosophy for the innumerable discussions concerning problems related to the material in this book. Without these Thursday discussions, it would have been difficult to really learn the new way of thinking needed in these questions. I am especially grateful to Professor Jussi Rastas for his competent and careful criticism.

I wish to thank Virginia Nikkilä, B.A., for the revision of the English language in most chapters of this book and Karri Sunnarborg, M.A., for technical help in drawings. To my publisher, Springer-Verlag, I am grateful for keen interest, especially to Professor W. Beiglböck, Dr. Angela Lahee, Ms. Antje Endemann, and Ms. Ulrike Drechsler for many good proposals in the editorial work.

Klaukkala
July 1996

K.V. Laurikainen

Contents

Introduction

Seven decades ago, between the years 1924 and 1927, the most wonderful theory that the human intellect has created to date – excluding purely mathematical discoveries – was brought into being. Bohr's atomic model, where electrons were supposed to be moving around the atomic nucleus according to mechanical laws and were bound to the nucleus by an electric force, was found to be unsatisfactory in many respects. There was no doubt, it is true, that the road into the atomic world which Niels Bohr had found in 1913 was right, but this road had been followed to its end, as Bohr himself had shown in his 1922 lecture series in Göttingen, at the famous 'Bohr Festival'.

A promising new idea was found in 1924 by Louis de Broglie, when he, in his doctoral thesis, pointed out that the strange dualism characteristic of electromagnetic radiation could perhaps also apply to different kinds of particle rays, i.e., to the motion of particles. It is well known that light is a form of electromagnetic wave but, on the other hand, Max Planck and Albert Einstein showed at the beginning of this century that light is composed of small energy quanta, of a kind of light particle (photon) which has a certain momentum and energy just like a small body. Thus, light seems to have two different faces which appear according to the method which is used when investigating light phenomena. This dualistic nature is also characteristic of particle rays, i.e., of particle streams composed of fast-moving material particles. Due to the endeavors of Werner Heisenberg, Erwin Schrödinger, and many other atomic scientists, quantum mechanics was created: a theory which gives a consistent description of all objects of the atomic world describing them as *both* particles *and* waves, whichever fits the circumstances.

A difficulty remained in the physical interpretation of quantum mechanics. Close collaboration between mathematical theorizing and experimental observation had produced an excellent theory which was later verified in the most diverse microphysical problems. But what, indeed, are those 'matter waves' which are inseparably associated with the 'particles' of the microworld? The interpretation of matter waves became an especially difficult problem which actually has to be deliberated even today. Nature had guided physicists along an unexpected road, and people still have widely different opinions about the lesson nature teaches us here. The problem is whether we are able to correctly interpret the message of nature.

In his poetic narrative *Aniara*, the Swedish poet Harry Martinson tellingly describes the situation in physics in 1926:

Uppfinnaren var själv fullständigt slagen
den dag han fann att hälften av den mima
han funnit upp låg bortom analysen.
(*Martinson* 1956, the 9th song)

[The discoverer himself was totally petrified / when he found that one half of / what he had found was beyond analysis.]

In their intensive discussions, which continued uninterrupted over several months, Bohr and Heisenberg at last found a solution which is called the *Copenhagen interpretation* of quantum mechanics and which Bohr presented in September 1927 at a conference in Como, Italy. Wolfgang Pauli participated in an essential way in its formulation. Unexpectedly, the interpretation was generally accepted, although most physicists have never quite understood the philosophy behind it. The Copenhagen interpretation has, in any case, established the language which is generally used when quantum mechanics is applied. It is described here in broad outline in Chap. 2 "The Atoms Have the Floor."

Matter waves are interpreted as *probability waves*, and this causes difficulties with respect to realism. It turns out that the reason for these difficulties is that the influence of the observer on the experimental results cannot be eliminated. We can get exact knowledge about objects in the outer world only within a given conceptual framework (theory). Nature gives a definite answer only to a definite question, and this presupposes a certain system of concepts. A quite new, unexpected feature is that the properties of a microsystem which are found in different experiments often are mutually inconsistent. Bohr has called this feature, characteristic of the microworld, *complementarity*. The mutually inconsistent results of different experiments must be accepted as complementary descriptions of reality which itself remains abstract.

In his analysis of observations, Pauli came to the conclusion that *physics and psychology must be understood as complementary sciences which only together are able to describe essential features of reality*.

This is a new point of view, and this book tries to approach problems of reality according to this view. So far, the view has been excluded from physics because physicists – and, in general, philosophers as well – reject ideas which introduce 'irrational elements' into the scientific world view. Quantum mechanics seems to guide us, explicitly, to such a perspective. It presupposes the abandonment of the Cartesian dualism which is so deeply rooted in Western thought. There is the risk that the message of the atoms becomes distorted because of hidden presuppositions based on Cartesian dualism.

Physicists have tried to find the fundamental laws of nature by analyzing ever-smaller structures ever more exactly. Characteristic of the belief in a deterministic mechanism of the world are differential equations which describe 'infinitely small' changes. The experimental investigation of microphysics, however, has run into a discrete structure: atoms and quanta. The analysis of the implications of this discreteness has forced physicists to admit that a complete (deterministic) description of phenomena is not possible. This concerns the idea of causality and, therefore, the basic trends of modern Western thought.

In the Western world science has become a kind of surrogate for religion, and according to a rather widespread view, a scientific attitude presupposes the rejection of religion. This often appears as clear religious discrimination, which so far has not attracted enough attention.

One important aim of this book is to show that a critical analysis of the possibilities and limitations of science, in light of the 'epistemological lesson' of atomic research, leads to a profound re-evaluation of the relation between science and religion. This is a task which requires efforts both from representatives of science and from theologians. It is to be regretted that scientific education is generally given according to the conditions stipulated by scientism and materialism.

The weakened role of religion in society has created a moral crisis, the depth of which has not yet quite been realized. It is an expression of intellectual hubris that science has strongly contributed to this development; the most important dimensions of reality grow dim when the requirements of the intellect are disproportionately emphasized.

Hubris is always followed by punishment. I fear that the Western world will not awake from its materialistic errors without a really terrible lesson, the signs of which are already clearly in sight.

The attitude of physicists towards philosophy today is essentially different from that among the creators of quantum mechanics and the Copenhagen interpretation. When Bohr died in 1962 – Pauli had died four years earlier –, the criticism by the younger generation of the Copenhagen interpretation became the dominant feature in the discussion concerning the foundations of quantum theory. Physicists had obtained a good technical education in the spirit of realism, and they could not understand the philosophical problems which the 'founding fathers' saw in the foundations of this theory. In a letter, Pauli wrote that he belonged to a generation who saw really deep problems in modern physics but was not able to solve them. The younger generation has not seen these problems at all – because of its 'realistic' belief that physics describes an objective reality which is independent of any observers and observations. Today physicists transfer such problems to philosophers.

I understand Bohr well when, in his famous 'last interview' (see Chap. 7 in Part II), he said, "I think it would be reasonable to say that no man who is called a philosopher really understands what one means by the complementary description." Philosophers are not accustomed to listening to nature. They are too strongly bound to the philosophical patterns of thought. The deep philosophical problems which Pauli meant have not been discussed since. People do not seem to realize the existence of these problems, which concern the role of the psyche in physics and the role of physics in culture: questions of values and the basic belief which gives direction to research. Since the 1960s, the basic belief has been in materialism and in the unlimited rationality of the world. After the use of nuclear weapons in 1945, Pauli feared that this attitude would lead to a catastrophe.

Today people no longer understand the ontological problems of quantum theory. The Copenhagen interpretation is criticized in a way which shows that the philosophy of the 'founding fathers' is quite misunderstood. We really need natural philosophy in physics departments, but it should be independent of the philosophical prejudices of the professional philosophers. In many respects, complementarity and statistical causality open new perspectives which should be studied carefully. But this presupposes a new attitude: time for and interest in philosophy among physicists.

Without my study of Pauli's philosophy, I would not have been able to write this book. However, this is not a presentation of Pauli's philosophy but a description of my own views. In Parts I and II and in the first chapter of Part III (Chap. 9), I have tried to describe the main features in the philosophy behind the Copenhagen interpretation. What follows then are my personal views on the basis of this 'Copenhagen philosophy'. I have put a strong emphasis on religion because, according to my view, the relations between science and religion can be seen in a quite new light on this basis. Although Pauli has been my most important guide, the reader should understand that I only give a personal view of what Pauli's philosophy implies in these important questions.

I
Problems

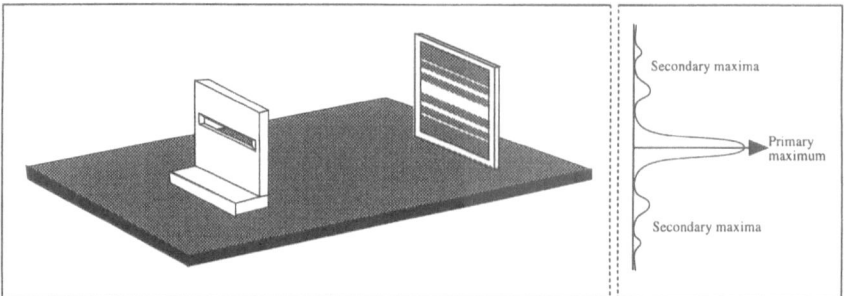

Statistical causality

Diffraction of radiation by a narrow slit. *Above*: Schematic drawing of the experimental arrangement. *Below*: In electron rays, the number of electrons passing through the slit can be regulated. With an increasing number of electrons the structure of the diffraction pattern becomes ever clearer. (Simulated results of a real experiment with two slits. See *New Scientist*, May 27, 1989, p. 39.)

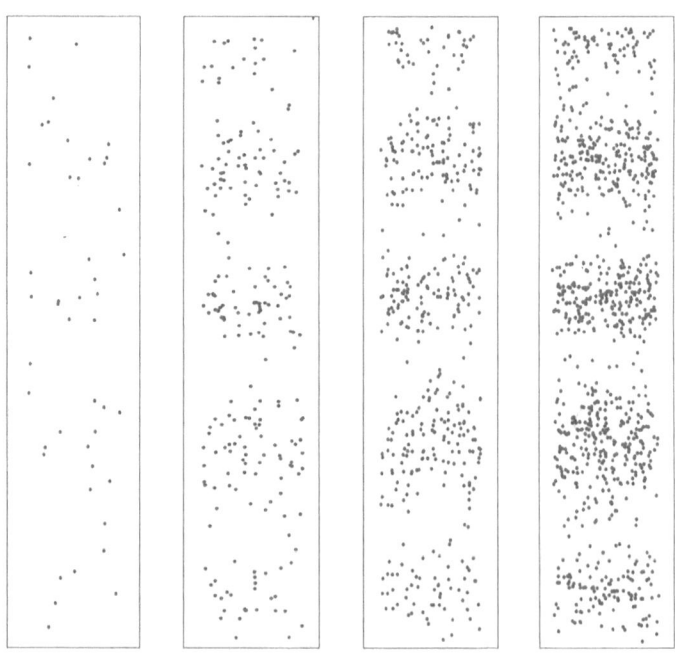

1. Purgatory

Usually I do not have any dreams. Instead, when I wake up I often have the feeling that I have been somewhere very far away where matters are understood more clearly than in a waking state. Often I feel as if I am awakening for a given task – in order to put down on paper something that was an idea in my mind when I regained consciousness. I have got the basic idea for much writing at just such a moment. It is like a given task, and when I begin to work, I can feel that I am almost like an instrument which writes down on paper ideas which actually are not my own at all. I have begun increasingly to write in the small hours, when I have the feeling that I am continuing to do what I was doing before I awoke.

I know where I was traveling this night, without dreaming about anything. I was with Dante Alighieri in hell and in purgatory. I have not had time to read of paradise, and it does not have any connection with this because I feel that I am in purgatory together with the whole of Western culture.

Mates in Purgatory

It is true that only very few people seem to have a similar feeling today. Some kind of foreboding of the realness of hell is generally gaining strength, however. People have a presentiment of where we are going, and this brings about worries and anxiety. But a clear view of the situation is lacking. The conception of reality is changing and therefore a firm moral basis has also been lost. Those in society who have the power and the obligation to state what is considered true now have a heavy responsibility indeed. They have to make it clear to themselves what is more important: truth or power. This is the decision between paradise and hell.

When market forces guide society, even the truth easily becomes a means that is evaluated using the measures of economy and power. "Knowledge is power," declared Francis Bacon. This is true today in a very concrete sense. Enormous sums, never seen before, have been invested in research, especially after the Second World War. Simultaneously, the evaluation of research has increasingly involved considerations concerning power.

Scientists themselves have taken an analogous attitude. The disinterested love of truth is beginning to be rare. It is also difficult to find acceptance for profoundly new views because the judgement of truth takes place in democratic circles and these are strongly bound to the customary lines of thought. Thus there is a strong temptation for a scientist to devote himself to questions where one can get intellectual and financial support. Nobody seems to be interested in questions concerning the foundations of knowledge – outside of philosophy, which has limitations of its own. For natural scientists, philosophy seems to be almost irritating. Therefore, the search for truth seems to be cast adrift: nobody tends the main course.

In this situation, who has responsibility for seeing that truth is not bound to the fetters of prejudices and the pursuit of power? Who owns the keys to making the distinction between truth and error?

The Church, and in particular academic theology, has a special responsibility. From that direction I would expressly expect companions in this purgatory where I am searching for clarity. I find it very alarming that religion – in such a form as churches represent it in the Western world – has difficulties with respect to educated people. This should create unprejudiced self-criticism in theologians.

The situation in the world requires open discussion now. In order for Christianity to maintain the role in the Western world which belongs to religion, its representatives must be ready to revise the language in which religious truths are expressed. *Religion and science must find each other again.* This is the most important task of ecumenism today. Science has become a religion in the Western world. This should be acknowledged by the representatives of the Christian theology, and they should try to achieve constructive communication also with representatives of this religion. Much will depend on the progress of this endeavor.

Dante in his time found a place for clergymen in the fires of hell. I do not dare, by any means, to begin to criticize the theologians of today in such a way, but my own feeling is that something essential is wrong. Some discussion of the relations between religion and science has been started in various quarters, but so far it has been just a preliminary sounding. One should overcome the discrepancy which has developed between religion and science during the last three hundred years and find a new harmonious relationship. If this does not succeed, society will be heading for inestimable controversies. Life and search for truth are without foundation if truth means two totally different matters.

The fact is that for a long time religion has not been the only official representative of truth in the Western world. On the contrary, quite generally people think that religion belongs to the past, and that in the future science alone will be responsible for truth. In that case, philosophy would have a great responsibility for the direction of science. Unfortunately, philosophers in the West are strongly influenced by a conception of science which puts a disproportional weight on logic, and philosophers in general have a negative attitude towards religion. This strengthens the *scientism* – a hubris of reason – which has gained a hold in the academic world.

If I, according to Dante's example, should begin to show the representatives of different cultural fields to a place in the fires of hell, I would first mention philosophers. Their responsibility for the main direction of thought in the academic world is great, and I feel that superficial things have a much stronger influence on their attitudes than the unprejudiced love of truth. If a philosopher wants to develop his academic career, he must bind his thinking too early to a given pattern of thought: He must show that he is a consistent representative of the doctrine he has chosen. Man, in fact, has sufficient maturity for such a choice only when his academic career is already coming to an end. By then, a philosopher's hands are usually strongly tied by his earlier writings. Nevertheless, I welcome philosophers to join me in purgatory. Perhaps even they still have some hope.

A third group from which I would expect companions is the representatives of the state. The feeling is very strong that we are in a deep political crisis. The whole system which is used for taking care of social affairs is becoming unsuited to its purpose. The traditional parties do not correspond to the requirements of today. The continually growing environmental problems, the controversies between rich and poor countries and corresponding controversies within a country, the potential risk of nuclear weapons, new moral problems created by science and its applications – all this belongs to the politics of today and even more those of tomorrow. New ideas are needed, as are a new philosophy and a new basis for morality. The power of materialism has clearly become exhausted, but it still lives on in the disproportionate adoration of market forces, and this guides development into a dangerously one-sided direction. Spiritual ideals and values are now needed instead.

Without religion, ideals and morals do not have strength. Is Christianity able to meet this challenge? That is the question.

The *arts* can in a way similar to religion influence people's functioning and decisions. They can also reveal truth. In this sense the arts have a real, vital force – and in this respect are not far from religion.

A Forbidden Perspective

Empirical science changed our conception of the basic nature of the world so that people began to consider the world as a deterministic machine where all changes were governed by absolute natural laws. Everything that happened was in principle predetermined by laws. Every event had its 'natural causes' from which the event necessarily followed. All 'supernatural' occurrences were abolished from this world view. For the theologians, this was embarrassing, and the matter could be settled only by accepting Cartesian dualism: The material world and the spiritual world were considered in principle as totally separate. It is true that the picture of the world became less clear in this way, but a state of peaceful coexistence was reached between theology and natural science. According to Pauli, this means that a theologian and a scientist, when they meet, keep a definite distance from each other, bow repeatedly, and express mutual assurances that they have nothing to say to one another.

Atomic research has changed the situation. *Determinism has collapsed in physics* – and thus one has reason to wonder whether such a utopian idea of the laws governing occurrences can be reliable in any other field either. Simultaneously the border between 'natural' and 'supernatural' disappears, because there are no occurrences which could be claimed to have only natural causes. Something that can be called 'supernatural' can be found in every phenomenon. In atomic theory, deterministic causality has been replaced by the idea of *statistical causality*, and this means that individual events always contain something 'incomprehensible' or 'irrational'. (See frontispiece to Part I.)

Most physicists don't wish to think of the ontological implications of the new conception of natural laws. Niels Bohr himself refused to speak of the basic nature of the atomic world. He considered talk of 'reality' to be unclear: it cannot be used in scientific discussion. Those who are less willing to accept a statement that so strongly resembles positivism – or pragmatism – have for more than thirty years now made attempts to find some alternative way to describe the basic nature of the world other than by saying that there is something 'irrational' in each individual event. Philosophers have wholeheartedly participated in these attempts.

All such attempts have (more or less clearly) led into a blind alley, and therefore it is best to acknowledge the facts and state that determinism is dead and the border between 'natural' and 'supernatural' has disappeared. This forces us to consider the relation between knowledge and belief on a quite new basis.

It has been surprising to see how difficult it is to accept this fact. Even those who clearly understand that determinism is dead wish to maintain at least some glimpse of hope that perhaps it will in the future be possible to avoid the intolerable idea that there is something 'irrational' in nature: that everything cannot be explained by the methods of rational science. In my correspondence with many physicists who understand well the foundations of atomic theory, I have seen how difficult it is for most researchers to acknowledge the limits of science. It seems that this is caused by the repression of religion and mysticism. The physicist who has most clearly described the philosophical consequences of atomic theory recently is Bernard d'Espagnat, who has published many books and articles concerning these questions (e.g., *d'Espagnat* 1983a, 1993, and 1994).

Among the creators of the present atomic theory (quantum mechanics), only Wolfgang Pauli and Werner Heisenberg also tried to clear up the ontological implications (consequences concerning the basic nature of the world) of the theory. Pauli emphasized the *irrationality of reality*. When he tried to present his thoughts to other physicists, the resistance was so strong that he found the discussion of such questions hopeless. He repeatedly had similar dreams which he interpreted as a compensation for the repression which he himself pursued with respect to these ideas. When he died at the age of 58, his philosophical views had only preliminarily been published; one must search for completion from his extensive correspondence (see *Pauli Letter Collection*).

Without knowing Pauli's patterns of thought, d'Espagnat came to an analogous conception of the atomic world. His means of describing the situation is the idea of a *veiled reality*: Science can never reach all aspects of reality, not even approach it without limitations with its research methods. I pointed out in a letter to d'Espagnat that this view of reality is the same as that of Pauli. When Pauli speaks of the irrationality of reality, he means that because of this aspect we cannot completely describe reality by using the rational methods of science. d'Espagnat replied that he prefers not to speak of 'irrationality' because there are so few of us who understand the situation, and he would not like to endanger his relations with his colleagues. (I have received a similar answer from some other physicists.) d'Espagnat wrote that he wishes to use language which physicists generally use. His treatises are indeed written in the form of a very exact rational analysis; they fulfill very strict requirements of 'scientificality'.

My own experiences of attempts to describe Pauli's philosophy to my physics colleagues (or to philosophers) are depressing. However, this concerns such important philosophical insights that I feel it my duty to attempt to do so, after having worked my way into Pauli's world of ideas for some 20 years. If physicists do not wish to understand, perhaps there are other thinkers who are interested in the problems of reality and are able to think of the philosophical implications of quantum theory without materialistic prejudices.

Therefore I have persistently tried to illuminate the ontological implications of quantum mechanics in my writings (mostly in Finnish). So far, this has resulted partly in certain misunderstandings. This is not unexpected because a very profound change in the patterns of thought is needed. But it has also been delightful to see that for many people the perspective which I have tried to illuminate has been a redeeming and clarifying experience similar to mine.

Truth and Freedom

If the boundary between 'natural' and 'supernatural' disappears, it also means that there is no clear boundary between religion and science. For a rationalist, this is an intolerable idea. On the basis of the decline of determinism, one can, however, bring forward very convincing arguments for such claims, and if something is considered intolerable, it does not guarantee that it cannot be the truth.

The breaking of the boundary between natural and supernatural does not mean that we should compromise the strict requirements concerning scientific knowledge. In this respect, people often misunderstand the situation. We should not compromise the requirements of the logical clarity which have produced the most beautiful results of science. Both Pauli and d'Espagnat, whom I have mentioned above as representatives of the new conception of reality, are simultaneously pure-blooded proponents of strict logic in their research. Their criticism is just more profound than that of physicists in general, and therefore they have understood that even the most rigorous empirical knowledge is not free of elements which have the nature of belief and which in fact give direction to the whole endeavor.

It must be emphasized over and over again that the new views of knowledge and reality are based on a careful analysis concerning observations which atomic research has made necessary. It is true that today only a few physicists think of the problems concerning the nature of reality and knowledge. In general, physicists refuse to include psychic matters in their considerations. They think that such matters are not of any interest in physics. But on what grounds?

Physicists are accustomed to understanding everything on the basis of the materialistic conception of reality, and when they meet expressions of the *consciousness* – and this cannot be avoided in an unprejudiced analysis of the observations – they feel that they have come to a limit which physicists may not exceed. However, a simple picture can be attained in the observation problems of atomic physics only if one states that the quantum mechanical description concerns *our knowledge* of the situation, not the atomic world itself. The theory presupposes implicitly 'that which knows' – the consciousness – and here one meets something that cannot be described by the methods of physics.

Thus, in the most rigorous empirical science we clearly meet the limits of the scientific methods. In fact, the situation is the same in all empirical sciences – sciences based on observations. However, in atomic physics the problems of observation have required an especially critical analysis. The result means the end of materialism as a scientific conception of reality.

Everything is reduced to the fact that natural laws do not correspond to the idea of strict determinism. One meets the 'irrationality of reality' or the fact that details of reality remain 'behind a veil' because of the nature of phenomena. Obviously, science can never reach an objective reality. What then is the truth? Even statements that cannot be explained by the 'correct use of reason' – on rational grounds – can be true. Here we meet a profound question concerning human knowledge and the picture of the world. If we try to be unprejudiced, we see that this question, which is met on the basis of science, cannot be strictly distinguished from the question of what truth means in religion.

I know that here a chorus replies from all sides that this is senseless. I am breaking the boundaries between categories, and then all kinds of absurdities follow and a complete muddle results. But this is exactly the reason why I feel that I am in purgatory. What is the truth? This is a great question which must be answered – or at least one should find some clarification for it in this new situation. In the 17th century, the same question was asked, and one found an answer to it. This answer has given direction to science and to Western thought in general. It implied that the world of material phenomena can be separated from the world of the spirit. The result was the 'spiritual cloud of fog' of the Western culture represented by psychophysical parallelism and the enormous strengthening of materialism.

Now nature has taught a new lesson to physicists (to those of them who will listen). It tells them that the world of matter cannot be separated from the world of spirit. There is only one world which appears to us in some connections as material, in other connections as psychic. In physics this also appears in the dualistic nature of the atomic world: its objects are in some connections material particles, in some other situations waves. The concept of object dissolves in the hands of a physicist. An object exists only with a certain probability, and its properties depend on the way it is observed.

But the same problem appears in another form as well. Natural laws are not absolute but probabilistic. Then each individual event implies a choice between the different possibilities that the probabilistic law allows. These *free choices* are expressions of the 'irrationality' which is the new property of reality. These choices are really free; they are not bound by any rational laws. One cannot explain them with the aid of rational science.

This is the new situation which we must acknowledge when we begin to search for an answer to the question, *what is the truth?* We must acknowledge that there are facts which cannot be explained by rational science and that there are really free choices. Those who fully understand this situation begin to see that the main direction of our culture is changing. One has to find a new answer to the question of what is the truth. Science alone cannot find an answer, nor can religion by itself. A dialogue is needed. And as one can understand, the foundations of morality are also at stake. Choices are based on values, and these are also the basis of ethics.

It is nothing new that these questions belong to the sphere of purgatory. In the second book of Dante, in the sixteenth song, the Lombardian Marcus, whom the author meets when climbing the mountain of purgatory, answers Dante's question of why there is so much evil in the world in the following way:

Then heaving forth a deep and audible sigh,
"Brother!" he thus began, "the world is blind;
And thou in truth comest from it. Ye, who live,
Do so each cause refer to Heaven above,
E'en as its motion, of necessity,
Drew with it all that moves. If this were so,
Free choice in you were none; nor justice would
There should be joy for virtue, woe for ill.
Your movements have their primal bent from Heaven;
Not all: yet said I all; what then ensues?
Light have ye still to follow evil or good,
And of the will free power, which, if it stand
Firm and unwearied in Heaven's first assay,
Conquers at last, so it be cherish'd well,
Triumphant over all. To mightier force,
To better nature subject, ye abide
Free, not constrain'd by that which forms in you
The reasoning mind uninfluenced of the stars.
If then the present race of mankind err,
Seek in yourselves the cause, and find it there;
Herein thou shalt confess me no false spy.

"Forth from His plastic hand, who charm'd beholds
Her image ere she yet exist, the soul
Comes like a babe, that wantons sportively,[2]
Weeping and laughing in its wayward moods;
As artless, and as ignorant of aught,
Save that her Maker being one who dwells
With gladness ever, willingly she turns
To whate'er yields her joy. Of some slight good
The flavour soon she tastes; and, snared by that,
With fondness she pursues it; if no guide,
Recal, no rein direct her wandering course.
Hence it behoved, the law should be a curb;
A sovereign hence behoved, whose piercing view
Might mark at least the fortress and main tower
Of the true city. Laws indeed there are:
But who is he observes them? None; not he,
Who goes before, the shepherd of the flock,
Who chews the cud but doth not cleave the hoof.[3]
Therefore the multitude, who see their guide
Strike at the very good they covet most,
Feed there and look no further. Thus the cause
Is not corrupted nature in yourselves,
But ill-conducting, that hath turn'd the world
To evil ...

---

[1] The human soul is freely subject to her Creator. Albertus Magnus holds that there is in man a twofold principle of action, nature and the will. Nature, indeed, is governed by the stars, but the will is free. Notwithstanding this freedom, however, the will will be drawn and inclined by nature, unless it steadfastly resists; and, since nature moves with the movements of the stars, the will then, if it does not resist, begins to be inclined to these stellar movements.

[2] Cary aptly notes: "This reminds us of the Emperor Hadrian's verses to his departing soul: Animula, vagula, blandula, etc."

[3] Cf. Lev. XI. 4. What Dante says is that no man attends to the observance of the laws, *because* the Pope can chew the cud (i.e., mediate and understand the Scriptures), but does not divide the hoof (i.e., confuses the spiritual with the temporal power).

(Translation by Henry Francis Cary, published in 1814; new edition by Edmund Gardner in 1908, with new commentaries, Dent & Sons, London, repr. 1961.)

2. The Atoms Have the Floor

The Burden of Rationalism

While wandering in the mountains of purgatory, Dante met souls which carried such heavy burdens that their heads were always weighed down, toward the earth. So, too, are the rationalists of modern science and analytical philosophy, who carry such heavy requirements of exactness and of the 'scientific method' that they cannot see anything else than the world of matter.

I have tried to point out to my colleagues representing theoretical physics that the 'paradoxes' which burden quantum mechanics disappear if we take into account that the state function describes *our knowledge* of the situation. The reduction or 'collapse' of the state function because of observation simply means that we 'become conscious' of certain new facts, and therefore our knowledge of the situation changes. This remark is refuted, however, by pointing out that a physicist cannot speak of 'consciousness' without defining what 'consciousness' means. The definition must, of course, fulfill the normal requirements of accuracy which characterize the exact sciences. Perhaps one should introduce a mathematical symbol representing 'consciousness' and then operate with it according to certain exact rules! When I have remarked that 'consciousness' cannot be defined in the same way as physical concepts, the reply has been that then it is best to continue the discussion of the foundations of quantum mechanics on the exact basis which has been developed during the last thirty years. Indeed, theoreticians and philosophers have a lot to do!

When I have tried to point out the irrationality of reality, a professor of philosophy replies that he does not see any need to imagine that reality would have other aspects besides rational ones. In the description of individual events, there are gaps, of course, and gaps can be expressions of a fundamental indeterminism of events, but it is not necessary nor even allowed to place anything in the gaps. "Although *there is a hole* in a sock, it is not necessary to suppose that *there is something in the hole* – for example some hidden spirit complementary to matter or an 'irrational factor'" (*Niiniluoto* 1984, p. 128).

So it is. If one does not realize that there is something else besides handtouchable matter, it is not necessary to pay special attention to the existence of the 'gap'. Anyhow, the 'gap' in the description of the individual events was the matter which caused the famous disagreement between Bohr and Einstein concerning the 'completeness' of quantum mechanics. The very fact that quantum mechanics is a statistical theory – that natural laws are probabilistic and do not correspond to the idea of determinism – was the heart of Einstein's criticism of quantum theory, which he repeated time and again. It has caused a very comprehensive body of literature concerning the foundations of quantum mechanics and also the famous *EPR experiments* (EPR = Einstein, Podolsky, and Rosen). However, many (perhaps most) philosophers and physicists do not find these issues truly interesting. But precisely in them hides the most important philosophical gift of quantum mechanics.

Physics and Metaphysics

Characteristic of rationalists is an aversion to metaphysics. Physicists in general belong to these rationalists, who have an unshaken belief in the possibility of deciding with the aid of science all meaningful questions concerning reality. Usually physicists even believe that all important facts can be described by using mathematics. This was also Einstein's belief.

The probabilistic laws contain a severe problem from the point of view of the rationalists: In the individual events there always seems to be something that cannot be governed by theory. This is the simple fact which shows that a rationalist's belief in the power of reason is unfounded. When one thinks more carefully about this issue, one is forced to admit that simultaneously the borderline between physics and metaphysics becomes unclear.

If the world is such that it is not possible to govern individual events, even in principle, then it is impossible to give complete 'natural causes' for a given phenomenon from which this phenomenon would follow. Here we meet, in its very simplicity, the *irrationality of reality*, which the rationalists cannot, by any means, accept into their picture of the world. They are ready to pay a very high price indeed for this belief of theirs, because the foundations of quantum mechanics become twisted in a very complicated way if one wishes to rescue the rationalistic belief. But in a welfare state everything is possible: Such attempts are willingly paid for with research funds and at international conferences they sell well. The 'irrationalists', instead, do not get the floor at the physics conferences because physicists are pragmatists and like to concentrate on more concrete problems (which give results more easily).

In order to understand the situation in this essential question, we must think about it in a little more detail because this book is explicitly based on the facts met when investigating the atomic world, especially the change in the conception of causality. Of the founding fathers of quantum mechanics, only Pauli has clearly presented the ontological implications (implications concerning the basic nature of the world) of the probabilistic laws.

It is important that Pauli describes the probabilistic laws of atomic physics as 'primary' – today one usually speaks of *genuinely* probabilistic laws – because in this respect the statistical causality of quantum mechanics differs from the statistical laws of classical physics, which are used in thermodynamics, for example. In atomic physics, it is not a question of imperfect knowledge but of the very nature of the phenomena.

This judgement concerning the situation is characteristic of the Copenhagen interpretation, and in this respect all the most important representatives of the 'Copenhagen school' had the same opinion – I mean here, especially, Bohr, Heisenberg, Pauli (Hamburg), and Born (Göttingen). The terminology used in the applications of quantum mechanics is based on the Copenhagen interpretation, and over the decades it has turned out to be well suited to its purpose. Because of this we can speak of a new conception of causality. All 'Copenhagenians' understood that this change concerns causality in general, not only the laws of atomic physics. Bohr, Heisenberg, Pauli, and Born described these matters in different ways but the main point was the same: The idea of a deterministic causality must be abandoned. I think that the term *statistical causality* expresses best the new conception of natural laws characteristic of the Copenhagen interpretation.

Pauli, in addition, emphasized the importance of psychology. His ontological remarks have therefore been partly misunderstood, all the more so since *the unconscious* is very important for him and he understands it in the sense of Jungian psychology, not in Freud's sense, which is better known.

Many people have found that Pauli became very 'mystical' in his last years, and therefore his philosophical ideas have not been taken seriously, even if people have heard of them. However, Pauli was and remained deep in his heart a rationalist in his scientific work. But when concentrating on the foundations of quantum mechanics, Pauli found that it is not possible to omit psychological aspects in analyzing the nature of observations. Then he also saw that there is no absolute border between the scientific conception of reality and mysticism. This was the result of a most rigorous criticism, not an expression of a 'weakness for mysticism'.

Simultaneously, the border between physics and metaphysics becomes dim. Besides a strictly rational science we must – especially during the phases of a strong development of science and when examining its philosophical foundations – accept knowledge of another kind, which perhaps can best be called intuitive. It cannot be described by using exact concepts and strictly rational theories. We must also accept less exact means of describing reality – for example, the use of analogies.

The Irrationality of Reality

The decline of determinism in atomic physics necessitates a new conception of the nature of natural laws. *It is possible to find 'laws' for the statistical mean values of physical quantities, but not for their values in individual events.* The Copenhagen interpretation of quantum mechanics expressly presupposes the abandoning of deterministic causality. This must be considered a fundamental experimental result which puts limits on the rational description of phenomena. This description cannot concern individual events but only the mean course of events.

It is exactly the non-anticipatable scattering of individual events around the mean course that Pauli emphasizes as an expression of the irrationality of reality. (See frontispiece to Part I.) The existence of natural laws shows that the rational description of phenomena is possible. It is characteristic of modern Western thought that this rationality is considered a necessary property of reality: Everything that is real is supposed to be rational. Pauli criticizes this belief, however, and calls it the *repression of the irrational*. He points out that the belief in complete rationality has become untenable because of the decline of determinism in physics.

The Western belief in rationality has a long history. When Plato held the world of ideas as 'that which truly is', he expressed this rational conception of reality. Analogously, Descartes presupposed in his dualistic picture of the world that reality is rational by its very nature. This presupposition has become ever more characteristic of science since the 17th century, and now also dominates the main course of the research into the foundations of physics. Pauli, however, interpreted the 'epistemological lesson of atomic physics' – the decline of determinism – as an extremely important experimental result: It forces us to abandon the strong belief in the rationality of reality.

The usual objection is that even statistics and the calculus of probabilities are rational and, thus, statistical laws do not presuppose any irrationality. In these cases, however, rationality concerns only theory and theoretical predictions. Irrationality, instead, is met when one compares theory with experimental results. It is a property of reality. The irrationality of reality means that it is wrong to assume that any rational description would reach the 'reality itself'. In fact, it is even wrong to expect that the scientific description of reality would without limit (asymptotically) approach reality.

In the light of quantum theory, the 'reality itself' must be seen as *veiled* (d'Espagnat). We use this term here for describing reality in a world where causality is not deterministic but statistical. The irrationality of reality associated with indeterminism forms a 'veil' which always makes a rational description of reality incomplete. (d'Espagnat justifies the idea of a 'veil' by saying that reaching reality itself would presuppose superhuman abilities.)

The most important trends in modern philosophy require such high conceptual accuracy that the discussion of the most important questions of existence becomes impossible. The 'irrationality' of reality opens questions which cannot be described by using very exact concepts, and therefore people may refuse to speak of such questions in science. This seems to be the most important reason why many people find it very difficult to accept Pauli's conception of reality, although it is a necessary consequence of the Copenhagen interpretation.

Often people remark that perhaps the future development of physics can open the possibility of describing even individual events uniquely. The

Copenhagen interpretation, however, presupposes that the quantum mechanical description of atomic phenomena, which is based on probabilistic laws, is as complete as the nature of these phenomena allows, and therefore the description of the individual events is in principle impossible: This claim corresponds to the basic nature of the atomic phenomena. The application of quantum mechanics based on this philosophy has continuously produced excellent results, while Einstein's attempts to describe microphenomena more completely have failed. The great majority of physicists have, indeed, accepted the view that natural laws are genuinely probabilistic, and this implies necessarily the irrationality of reality.

On the Concept of Irrationality

Because the terms *rational* and *irrational* seem to cause misunderstandings, I shall try to make some clarifying remarks. The word 'irrational' is often understood in a quite different sense (for example, to mean 'contrary to reason'). Therefore, it must be emphasized that in the following irrational means the opposite of rational – as in mathematics one speaks of rational numbers and irrational numbers. The irrationality of reality means that rational science is not able to describe everything that is real. It is not possible to define exactly matters which are irrational. For example, an individual event cannot be described in a unique way if causality is not strict (deterministic) but genuinely statistical.

Irrational, thus, means the same here as *non-rational*. Therefore, one should clear up what the term 'rational' means. This cannot be done in a unique way, however. Pauli used the term rational in a very strict sense. For him only descriptions which belong to some logically correct – in general, mathematical – theory were rational. Which matters can in this sense be described in a rational way remains basically an open question, but the Copenhagen interpretation is based on the definite view that individual events do not have any complete rational description or explanation because laws in atomic physics are *genuinely* probabilistic.

In this connection, it is appropriate to think of the different kinds of knowledge in Plato's philosophy. Plato uses the term *doxa* for the lowest kind of knowledge, which can be described as 'beliefs' or 'opinions'; these are statements which we use in non-scientific everyday language. Scientific language is called *dianoia*. As a typical example, Plato mentions geometry with its 'eternal' truths. This kind of knowledge corresponds to rational knowledge in Pauli's terminology.

In addition, Plato speaks of a third kind of knowledge, which is the highest of all. Its name is *episteme*, and it means the comprehension of ideas which concern 'that which truly is'. We could call this kind of knowledge metaphysical. It is not rational according to the terminology which we are now using, although Plato held it in especially high esteem.

Pauli also appreciated metaphysics very much. He considered it the source of truly new scientific ideas, and therefore he warned of drawing a strict border between science and metaphysics. One should not define what is 'rational' too strictly. What today is 'irrational' can perhaps be described with the aid of rational theories in the future.

The appearance of the finite quantum of action (Planck's constant h) in the atomic theory brings into this theory a discontinuous (steplike) element which causes a new situation in the theory: The concept of causality is changed. The Copenhagen interpretation of quantum mechanics, accordingly, presupposes that we must definitely abandon absolute causality (determinism) and adopt the idea of statistical causality. The rational description of individual events then becomes impossible.

One could hope, of course, that the development of physics would lead to a new theory of microphysics which is strictly causal. Then one should abandon the basic ideas of the Copenhagen interpretation. Until now, however, everything has pointed in the opposite direction: We have reason to think that in the future physics will be driven still further away from the utopian idea of determinism.

In individual events we meet the irrationality of reality. There are aspects in reality which cannot be described by the rational methods of science. This means that we have opened the door to a new world.

Consciousness Enters into Physics

Since the representatives of the Copenhagen school realized that they could not describe reality itself within their atomic theory, they had to state that quantum mechanics describes only our knowledge of reality. This is an extremely important statement which has appeared in the writings of the creators of quantum mechanics at least since the 1950s. In quantum mechanics one has landed in a situation where theory does not describe reality itself but nonetheless offers very exact knowledge of real events – although only probable knowledge, it is true. Thus, one cannot verify the truth of the theory by comparing it to reality itself, because this is and will remain 'behind a veil'.

If one speaks of 'our knowledge', this presupposes that there is something or somebody 'who knows'. Quantum mechanics brings the *consciousness* into physics. This means the end of materialism in physics. Quantum mechanics is a complete description of the material world on the level which this description concerns. The wave function (or the state function) is as complete a description as the nature of matters allows of the state of the system in the situation which has been defined by the experimental arrangements. However, if a satisfactory picture of the whole situation is wanted, besides this description concerning the material world, one has to acknowledge the consciousness of the observer as an element of reality of a totally different kind which can scarcely be described with the aid of quantum mechanics.

Pauli has strongly emphasized that an observation presupposes, in fact, some kind of 'interaction' between the object of observation and the consciousness of the observer. Bohr, instead, wanted to guarantee the objectivity of science to such an extent that he excluded all psychic aspects from his considerations. Then he was forced to emphasize that the objectivity of an experiment presupposes an accurate description of the method which was used. The choice of the experimental method presupposes psychic actions. Bohr tried, however, to describe quantum mechanical experiments without any mention of psychic matters. This is an unnatural limitation, which Pauli tried to point out to Bohr in the middle of the 1950s, but Bohr did not change his view. Bohr considered an observation a purely physical interaction between the object of the observation and the measuring apparatus. He understood an observation to be finished when there was a 'result' in the registering instrument.

In two respects we must criticize this theory of observations which Bohr presented in the 1950s and from which psyche has been completely excluded. First, according to Bohr's account, 'objectivity' means the same as *intersubjectivity.* It does not mean that our picture corresponds to reality itself, but only presupposes that experts have an identical opinion of the result; d'Espagnat speaks in such a case of a *weak objectivity.* Bohr's conception of 'objectivity' implicitly includes the activity of the human psyche.

Second, it must be emphasized that the observer's psyche is active in the accomplishment of the experiment, first in planning the experiment and finally in the interpretation of the results. It is quite arbitrary to not mention this only in order to avoid the introduction of subjective elements into physics. The 'result' which is safe in the registering instrument – for example, a set of zeros and ones on a magnetic tape – does not mean any knowledge concerning atomic events before this 'result' has been interpreted according to a given theory.

When Bohr emphasized that even in quantum mechanics the observer can be considered a 'detached observer' quite in the same sense as in classical physics, Pauli remarked that this is in contradiction to Bohr's often repeated statement that "we are not only spectators but simultaneously actors in the great drama of existence". When choosing the experimental method, the observer influences the results of the observation: whether he will see particle trajectories or interference patterns of waves, or correspondingly, whether he gets an exact value for position and an inexact value for momentum or vice versa, etc.

Bohr took refuge in emergency solutions, in which psychic matters are completely excluded from physics, because he feared that otherwise we would be leading science towards mysticism, which he considered highly undesirable. Pauli has, on good grounds, pointed out that this is against the most characteristic features of quantum mechanics. We cannot ignore the fact that the free will of the observer has an influence on the results – because the results depend crucially on the method of observation – and that consciousness influences the interpretation of the results.

Complementarity

Niels Bohr used the concept of complementarity when characterizing the nature of atomic physics. Quantum mechanics substitutes operators for the quantities of classical mechanics. These are often defined as pairs of operators which exclude each other: The corresponding quantities, which are called complementary, cannot simultaneously have exact values. Such operators are defined simultaneously, using an algebraic relation which contains in a symmetric way both operators of the pair. A standard example of such a pair is a position coordinate and the corresponding momentum component. The more accurately one quantity of a complementary pair is measured, the more inaccurate becomes the value of the other quantity. The *uncertainty relations* of Heisenberg give an exact limit for the accuracy.

In this sense quantum mechanics is complementary by nature. Operators, which here replace the quantities of classical mechanics, are not measurable 'quantities' in the usual sense but are very abstract and describe symbolically 'something' that cannot be visualized but only described mathematically. The objects of quantum mechanics also remain very abstract; they manifest themselves in some experiments as particles, in others as waves. Objects and their properties can be understood in quantum mechanics only as certain metaphors of reality. Pauli - like Bohr - said that reality can only be described symbolically.

Complementarity can be held as the most characteristic feature of quantum reality (as Bohr did). When investigating complementary properties of reality, we need measuring methods which exclude each other. The observer has to choose the method, and the results depend on this choice. Therefore, the existence of an objective reality, which would be independent of the observers and their observations, becomes problematic. This is characteristic of Bohr's philosophy of complementarity.

All representatives of the Copenhagen school (especially Heisenberg, Pauli, and Born) adopted Bohr's philosophy of complementarity. Heisenberg liked to emphasize his uncertainty principle, while Pauli as well as Born described statistical laws (statistical causality) as the most characteristic feature of the theory. However, all of them accepted Bohr's complementarity as a pertinent description of atomic phenomena.

The Psychophysical Problem

The most important members of the Copenhagen school – perhaps excluding Bohr himself – have made remarks which show that complementarity also opens a new perspective on the relation between matter and psyche, i.e., the psychophysical problem. All of them criticized Cartesian dualism. These ideas were developed in their most detailed form by Pauli in his correspondence with Markus Fierz and C.G. Jung. Pauli held the psychophysical problem – and the question of a new conception of reality – to be the most important problem of our time.

This is characteristic of Pauli's philosophy. For Pauli, reality means a whole which includes both the 'outer world' and the 'inner world'. A physical observation is always an 'interaction' between an object belonging to the outer world and the consciousness of the observer. However, this 'interaction' cannot be considered similar to the causal interactions between the objects of the outer world. It is an 'interaction' of a kind which is quite unknown to contemporary science. The signals which arrive from the outer world start processes in our unconscious psyche, and the result of such an unconscious process is that we 'become conscious of something' – and this is an observation.

It is very important to understand that factually 'an object of the outer world' is a picture in our consciousness. This picture is a result of a complicated process which has taken place in the unconscious; we can never analyze this process in detail because it is unconscious. This alone makes it understandable that there is a 'veil' which hides 'reality itself' from our consciousness.

It is characteristic of Pauli's thought that he strongly emphasized this role of the unconscious in the observation process which is the basis of empirical knowledge. He was especially interested in the concept of *archetype*. This is the central theme in Pauli's extensive Kepler article in the volume *Naturerklärung und Psyche*, which Pauli published together with C.G. Jung (*Jung & Pauli* 1952). Archetypes were of mutual interest to the two men.

Archetypes are inherited or learned basic types of the shaping (gestalting) activity of the unconscious – 'instincts of imagination'. Pauli describes them as a bridge between our sense perceptions and ideas (concepts). Archetypes belong as much to the outer world as to the inner world. These two 'worlds' are complementary expressions of 'reality itself'. These complementary elements of reality are inseparable. We can get a reliable picture of reality only by considering it as a whole containing both matter and psyche.

Pauli was a realist, since he especially emphasized the importance of searching for a new conception of reality which is compatible with atomic physics. However, his realism was quite different from a realism which tries to describe the outer world without any subjective elements of psychic origin. Pauli understood realism in a deeper sense: He stated that the objective description of the outer world is impossible because our conception of the outer world is a result of our psychic shaping activity, and the psychic processes in the unconscious are an inseparable part both in observations and in theory construction.

Instead of Cartesian dualism we need the conception of complementarity in the description of the relation between psyche and matter. We experience the inner world as psychic phenomena and the outer world as material phenomena, but everything that we experience belongs, in fact, to our consciousness. These two 'worlds' are basic features of reality, which is neither psychic nor material but which we experience in these two complementary ways. According to Pauli, reality is abstract (non-visualizable) in the same sense as the atomic world.

The interaction between these two 'worlds' is a connection (regularity) which is fundamentally different from the interactions between the objects of the phenomenal world; it cannot be described by natural laws. If we try to describe this interaction, we can learn something from the complementary relations of quantum mechanics. But we must note that here we are describing processes which take place in the unconscious, and these cannot be described with the same logical accuracy as the processes of the phenomenal world. We are in the realm of metaphysics.

The complementarity of psyche and matter can be compared with the complementary properties of the objects of the atomic world; these appear in observations as 'particles' or as 'waves', depending on the observation method which we choose.

According to this view, physics and psychology are complementary sciences. Both are needed if we wish to form a reliable picture of reality. A struggle for realism concerning only the outer world is not true realism. Unfortunately this struggle dominates the basic research of quantum mechanics today, partly because Bohr himself wished to exclude psychic matters from physics in the name of the objectivity of science. However, realism means that we acknowledge the facts, and we cannot deny the fact that psychic activity is essential in observations.

3. Scientism

Towards a New Conception of Science

What, indeed, is at the heart of the message we have received from the atoms? Bohr, Heisenberg, and Pauli each emphasized different features when describing the 'epistemological lesson' given by atomic research. Examined more carefully, their basic views are closer to each other than they seem at first. Let us compare their views, because the differences between them are often exaggerated.

Bohr emphasized the idea of complementarity. He found it to be the essence of the 'strange situation' (*eigentümliche Lage*) that we encounter when describing atomic phenomena. The result has been an excellent theory with splendid experimental verification, but we cannot claim that we have learned to describe the atomic world. According to Bohr, quantum mechanics offers an extremely good description of atomic phenomena. Certainly, it is characteristic of *atomic phenomena* that experimental results can only be predicted in terms of probabilities, but even probabilistic predictions can be very reliable. A truly strange thing, however, is that atomic objects have complementary properties.

What is the basic nature of objects which in certain experiments are 'mass points' and in other experiments 'waves'? In a physicist's world of concepts, these have contradictory properties. The 'objects' of the atomic world are not objects in the same sense as one speaks of 'objects' according to the traditional realism. Therefore, Bohr refused to speak of 'reality' and of 'realism'. In addition, he chose for his coat-of-arms the Taoist Yin-Yang symbol – but thereafter he found it necessary to repeatedly emphasize that mysticism should have no place in science! This inner controversy of Bohr's has made his message unclear and has caused much confusion.

Heisenberg was fond of mathematical theories, and he especially emphasized the limit of accuracy which in quantum mechanics is characteristic of complementary quantities. A 'mass point' which has only an uncertain position and momentum – as described by Heisenberg's *uncertainty relations* – can be imagined to be, simultaneously, some kind of 'wave packet'. However, one should not associate precise images with these concepts because then one easily runs into difficulties. Quantum mechanics is an exact, mathematical theory which cannot be visualized but which gives as exact predictions for experimental results as the uncertainty relations allow.

According to Heisenberg, one can speak of 'reality' only in the sense of potentialities, i.e., as possibilities which can become 'actualized' with certain probabilities. If we are satisfied with this kind of reality where statements can be presented with certain probabilities, then we can speak of 'reality' even in quantum mechanics, but this reality contains an essential irrationality: In an actual experiment there are in general several 'potentialities' which can 'actualize', and it is not possible to say which one of them will do it. Heisenberg's reality is a *world of potentialities* – nothing can be predicted with certainty.

Pauli's way of describing the situation in quantum mechanics is the one that I find the clearest. He emphasized the necessary change in the conception of causality. One has to abandon the idea of deterministic causality and to accept the *idea of statistical causality*. This means that individual events remain without explanation in principle, and here, in this fact, we meet the irrationality of reality. We cannot describe the reality itself; we can only describe certain of its features which can be reached by the rational theory. We cannot compare our theory with the 'reality itself' which remains 'behind a veil' for us. We can, however, construct rational theories which help us to govern atomic events to the extent that this is possible with the aid of probabilistic laws.

What is the moral to be found behind these views? First of all, we see that we cannot reach reality with the aid of our rational theories. However, the basis of empirical research is the *belief* that there is something that can be called an *objective reality, independent of human observations*. In empirical science we trust that observations give us some knowledge concerning reality which is not just imagined by us – and we must construct our theory according to this reality. Otherwise it does not function. On the other hand, atomic research has made it quite clear that the picture of reality which we form by using the methods of natural science is not independent of the methods which have been used in our research. Therefore, *traditional realism which presupposes the possibility of comparing theories with an objective reality, independent of observations, is not possible.* (It is true that few physicists think of such difficulties.)

However, experience strengthens the belief that we gain some knowledge about reality with the aid of scientific methods. Science is able to give predictions which help us to reach the aims which we pursue. It helps us to live in the world in which we are living. It also helps us to understand this world and our role in it.

Bohr, Heisenberg, and Pauli had very similar views about the ontological situation in atomic physics. Although Bohr emphasized the reality of atoms on the basis of the experimental results in this century, he on the other hand said that one cannot speak of 'reality' or 'realism' in atomic physics. He wished to say that people in general associate with realism expectations which do not correspond to the picture of atomic systems which quantum mechanics gives. Analogously, the irrationality of reality which Pauli emphasizes means that we can describe only certain rational features of reality: It is possible to describe only the statistical mean course of atomic phenomena, while a detailed description of individual events cannot be reached. Therefore, reality remains 'behind a veil', as d'Espagnat describes the situation. The same is the case according to Heisenberg as well: The 'potential world' includes 'irrational' aspects in the sense that one cannot know which possibility will 'actualize' in each individual case.

Beside 'independent reality', d'Espagnat speaks of *empirical reality*, meaning the best scientific picture of reality which we have. It is a changing picture, of course, vindicated by the criteria of reliability which are used in science. According to our experience, this empirical reality is valuable to us and helps us to live in this world, although it cannot be considered a 'correct' picture of independent reality. The 'potential world' of Heisenberg is an example of 'empirical reality'. It means, in fact, the same as the description of atomic phenomena according to quantum mechanics. I suppose that Bohr and Pauli accepted Heisenberg's description without hesitation as one possible way to describe the situation in atomic physics, although Bohr never used it himself, as far as I know; Pauli used instead the Aristotelian terminology of Heisenberg in some cases. The 'actualization' of one 'potential' case in an observation contains the 'irrationality of reality' which belongs to Pauli's vocabulary.

If we try to use language which might make the matter clearer to theologians, we can say that 'independent reality' appears to be transcendent (not reachable by human reason), but *belief* in it is necessary in order for empirical research to be meaningful at all. In addition, it is important to state that very strong empirical reasons strengthen this belief because empirical science seems to be extremely successful. In this respect a physicist's conception of reality and of the nature of human knowledge differs from Kant's view, according to which the description of natural sciences relates only to the phenomenal world, not to 'reality itself'. The idea of 'veiled reality' does not mean that there is a 'noumenal world' which cannot be described at all in empirical science.

Pauli significantly clarified the ontology implied by quantum mechanics. He emphasized the psychic aspects of the empirical research but this has remained unnoticed in the philosophical considerations. A picture of the world is always only a picture in our consciousness. It is always a result of the functioning of the human psyche and gets its form from our psychic properties, as Kant pointed out. However, Pauli did not accept Kant's categories. Instead, he adopted the pattern of thought of C.G. Jung's depth psychology and, in particular, the *idea of archetypes* characteristic of it. He tried to form, on this basis, an idea of how the knowledge of the world is created in our consciousness. According to Pauli, the *unconscious* functioning of the psyche must necessarily be taken into consideration if we wish to form a reliable picture of reality.

Pauli's philosophy differs essentially from Bohr's thought in this respect. Bohr wished to avoid psychological considerations in physics because otherwise the objectivity of physics would be lost, and in his later years he particularly emphasized the objectivity of science. Pauli stated that Bohr was inconsistent in this point because, when describing complementarity, he had emphasized that "in the great drama of existence we are simultaneously actors and spectators." In this way Bohr wished to emphasize that the planning of the observation method has an essential influence on what kinds of properties the atomic systems have in our experiment.

Here exactly is the depth of Pauli's philosophy. While Bohr wished to describe an observation as a purely physical interaction between the object and the measuring apparatus, Pauli described the measuring apparatus as a prolongation of the observer's sense organs and the observation as an 'interaction' between the object and the consciousness of the observer. This 'interaction', however, is of a quite new kind and draws attention to the psychophysical problem, as was stated earlier.

Problems which arise if we regard the functioning of the human psyche will be approached from different points of view in the following. The picture of physics or of human knowledge in general will change in a profound way. This does not mean that we have to change the exact sciences into something other than what they have been. We only have to learn to see science in a new, more general perspective and to understand the limits which a realistic judgement of the nature of human knowledge imposes on science.

For a rationalist who believes that all problems concerning reality can be solved using scientific methods, there cannot be any irrationality in reality. Nor does one see freedom, the importance of which was emphasized above.

In my correspondence with many colleagues who understand well the philosophy of quantum mechanics, I have seen how difficult the idea of the irrationality of reality is for physicists in general. The indeterminism implied by quantum mechanics (the fact that laws of nature are genuinely probabilistic) is certainly quite clear for physicists, but in spite of this, people are not willing to speak of the irrationality of reality. Obviously, the word 'irrational' has a too strongly negative meaning.

However, I have decided to follow Pauli's example and to speak of the *irrationality of reality* because I think that this expression points out strongly enough – and in the end in a very appropriate way – the change in the conception of reality which is a consequence of the decline of determinism.

Another way to describe this change is to emphasize the significance of *free choices* in the picture of the world. This expression also seems to cause difficulties for physicists. For many rationalists the appearance of such an element of freedom in the picture of the world seems to be in some way intolerable. Although it is necessary to abandon determinism, one wishes to find some compromise so that speaking of quite free choices, without any motives, could be avoided. The idea that choices take place in nature without any rational reason seems to be quite unacceptable to many people. The hidden variable theories, where one attempts to amplify quantum theory by adding to it variables which in fact cannot be measured, are examples of this attitude.

The Rebellion of Scientism

In fact, in the research on the foundations of quantum mechanics, one has, since the 1960s, explicitly tried to approach the problems of atomic theory in such a way that it would be possible to avoid the idea of the irrationality of reality. The emphasis has been on the exact definition of concepts and the logical clarity of the theory, to such an extent that the very problems concerning complementarity disappear. All of these new approaches have problems of their own. One can with good reason endorse the remark made in 1987 in Joensuu by the late Rudolph Peierls, who represented the old 'Copenhagen school', that these new approaches are not needed because quantum mechanics already has a natural interpretation. This was born in close contact with the process of creating the theory itself, and it describes in the best way how this theory has to be associated with observations. It leads to certain problems concerning the conception of reality, but this belongs to the 'lesson' given by atomic research. It is an empirical result, and one should not try to nullify it with the aid of logical operations.

However, the attempts to bring quantum mechanics into harmony with traditional realism continue. One tries to describe reality in a completely rational way. The fact that all such attempts lead to difficulties in some respects and are more complicated than the original Copenhagen interpretation should already show that the direction of the endeavor is wrong. One tries to nullify the most valuable philosophical gift of quantum mechanics.

The objection of a rationalist is that the scientific attitude does not allow any irrational dimensions in the picture of the world. He says that the 'burden of evidence' belongs to a person who includes in his picture of the world something that cannot be justified by scientific methods. In fact, I have given such arguments, but a rationalist does not find them convincing. *Scientism* excludes such matters from the conception of reality. However, does not the 'burden of evidence' belong to a person who, on the basis of scientism, denies the 'reality' of certain important matters – for example, of free choices – which are obvious according to normal thinking?

Even the belief that there is no irrationality in reality is a faith. Persons who are strongly inclined to rationalism easily appropriate this faith to themselves, but it is no more scientifically justified than a decision of faith in general. It leads to a truncated conception of reality which denies the 'reality' of many most obviously 'real' matters like freedom of will, which is one of the most important facts in life.

The denial of the irrational dimensions of reality leads even scientific research into questions where the reliability of scientific methods can be questioned with good reason. An example is modern cosmology, which is favored by physicists and astronomers although its reliability can be strongly questioned on the basis of the remarks presented here.

An observation is always based on a belief in a certain kind of reality, and experiments are planned on the basis of this belief. On the same basis we also interpret our observation. The result is a picture of reality which in essential aspects reflects our beliefs.

The belief in the objectivity of science is strong, and on this basis physicists – in spite of the teachings of quantum mechanics – are inclined to construct theories where the use of scientific methods is no longer justified but instead one is moved into the realm of faith. We shall illuminate this with a brief consideration of modern cosmologies.

The Big Bang

According to the most-favored cosmological theory, the universe came into existence through the 'Big Bang', where its volume was initially extremely small and its temperature extremely high. From this state the expansion of the universe, which is still continuing, began. As a result of processes taking place according to the known laws of particle physics, the porridge of elementary particles produced more massive particles and further galaxies, stars with planets revolving around them, etc. From the point of view of the classical picture of the universe, this kind of vision is, of course, a quite reasonable although very bold generalization of the experience collected in observations on the surface of the tiny globe called Earth.

There has been obvious interest among the public at large in these theories concerning the beginning of the world, and this has been utilized by conspicuously marketing Big Bang theories in literature and popular lectures as well as in exhibitions and TV programs, etc.

This has caused objections by some religious groups. As far as I have observed, these popularizations often do contain anti-religious propaganda (perhaps unintentionally), recommending in jest new, 'scientific myths' in place of the old religious ones. However, the nature of a myth has then been totally misunderstood: A myth should not be judged as a scientific theory; a myth should have an influence on our life and decisions.

Taking into account the remarks concerning human knowledge made above on the basis of quantum theory, these 'scientific myths' concerning the birth of the universe clearly exceed the limits which a scientific theory can ever reach. The unannounced presupposition in these theories is the Einsteinian faith: the belief that the world can be comprehended by human reason, that everything which is real can be described with the aid of rational theories. However, this very belief has been shown to be untenable by atomic research. In spite of this, Big Bang theories are still being developed and marketed with great enthusiasm, which shows how strongly scientism guides research today.

The idea that events fulfill a certain purpose or that that which happens is directed towards a goal has always been strong in the history of ideas. The purposefulness which can be seen particularly in organic nature gives cause for such an idea. In the modern natural science, this kind of thought or *teleology* (Greek *telos* = goal) has been judged to be unscientific, however. In a deterministic picture of the world, there is no place for such a thought, indeed, in any other sense than that in the beginning the Creator created the world for a certain purpose. *As a consequence of the genuinely probabilistic laws, the idea of teleology has again become possible in the scientific picture of the world.*

This opportunity has not been taken into consideration in the natural sciences so far. Since the 1960s (after the deaths of Pauli and Bohr), people have tried to develop and apply quantum theory according to the materialistic conception of reality, as has been the case in physics in general. Thus it has been forgotten that quantum mechanics was born from problems concerning observation, and observation brings in the concept of consciousness. Therefore, Pauli came to search for a new solution to the psychophysical problem. The basic philosophy of the Copenhagen school implies that we must abandon Cartesian dualism where the world of spirit is considered in principle independent of the world of matter.

Such views presented by the creators of quantum mechanics have been forgotten under the dominance of scientism. This is very clear in cosmology, where quantum mechanics is applied in a way which, from the point of view of philosophy, is very superficial. The implications of the basic faith are obscured by the deceptive veil of rationality.

If one wishes to describe the first phases in the evolution of the Universe in the spirit of quantum theory, one should think of evolution as a *series of quantum processes*, occurring step by step. According to quantum hypothesis, each step is an indivisible whole where a rational analysis of the details is not possible because each step contains an irrational element (a 'choice'). Although quantum mechanics is applied in cosmological theories, it is not done in this sense, and thus the discontinuity characteristic of quantum theory is shrouded by a description of continuous change.

A correct quantum mechanical treatment of evolution presupposes a series of discrete, indivisible steps governed by probabilistic laws – a so-called stocastic process. Each step contains an '*irrational choice*' which cannot be predicted by any rational theory. It can direct the evolution in a decisive way. Some especially critical theorists – I shall mention only Pauli and d'Espagnat – have seen in these 'choices' a *teleological factor* which guides the evolution.

It is usual to justify the methods used in cosmological theories by referring to the enormous number of events when describing the evolution of the whole universe; therefore, statistical laws can be very exactly approximated by deterministic laws. Thus it is claimed that we are considering a limiting case where deterministic laws are valid in a very good approximation, and it is not necessary to speak of any 'discontinuous steps' and of any irrationality contained in them. On this basis, the influence of the teleological factor is excluded, and one speaks instead of *pure chance* which guides the evolution.

However, the teleological (irrational) factor can systematically influence the 'actualization processes' and totally muddle rational expectations. Only a belief in the complete rationality of the world justifies the exclusion of teleology from the picture of the world; it cannot have any scientific justification.

If one tries to get some picture of the 'first events', it is very bold to exclude the possibility that some kind of 'irrational guidance' would be an essential, directing factor. Creation is irrational by nature, an act of free will, and an attempt to describe the birth of the world with the aid of a purely rational theory must be considered erroneous on principle. This mistake is caused by the Cartesian (and Kantian) conception, according to which the 'world of matter' is a closed whole; when describing its events, it is not necessary nor even allowed to take into consideration the free, creative will which belongs to the 'world of spirit'.

When considering the further progress of evolution, a rational description becomes better justified, and therefore modern cosmologies, too, certainly have scientific justification. One should just not imagine that the theory is at all able to describe the 'first events'. Even in the further evolution, one has to remember that the 'teleological guidance' can change the direction of the events in a way which cannot be governed scientifically. This means acknowledging the limits of science. There is both a rationally describable and an irrational, creative element inherent in evolution. Scientism alone provides the justification for the exclusion of the creative element from the present theories of evolution.

Scientific methods are rational, and it is not possible to compromise on this requirement. However, this does not justify the belief that all matters can be settled by using scientific methods.

What does this mean with respect to cosmology? First of all, it means a more critical attitude with respect to trust in the possibilities of science. The requirement of experimental verification must be presented much more rigorously than in the present cosmology. In theoretical physics research in general, the construction of mathematical theories without sufficient verification has become too common. It is true that mathematics is able to guide physics research with a quite unbelievable effect. The theory of relativity, in particular, is an excellent example of a theory which was created almost purely with the aid of logical analysis – but one should not forget that it was based on such classical theories as Newton's mechanics and Maxwell's electrodynamics.

There is, however, an essential difference between the present attitude in theoretical physics and that at the beginning of this century with respect to the connection between theoretical and experimental research. It is characteristic, I think, that all truly profound steps were taken in physics during the first three decades of this century.

Perhaps the most important reason for this is that the analysis of the philosophical foundations is belittled in physics today. The 'rejection of the irrational' is so strong among physicists that it bars them from paying attention to the facts. I mean, especially, that psychic matters cannot be neglected in a science which is based on observations.

Present cosmological models are only models for a possible evolution of a *world that can be totally described by mathematical theories*. This world in which we human beings live is not such a world, however; phenomena in this world always contain irrational aspects which are neglected in present cosmologies. These models presuppose a mechanical world which can be described by mathematical equations, while freedom and creativity are excluded from it.

It is possible to accept the main features of both the physical theories concerning the evolution of the universe and the biological theory of evolution because the idea of evolution is well justified in both cases. However, in both cases the rejection of teleology must be disputed. First of all, in any case, one should prevent the use of these theories as weapons in atheistic propaganda, as they are actually used, just because of the rejection of teleology in them.

If we believe that there is purposefulness, a struggle for a goal, inherent in the events, this raises questions of quite a different kind from those the present science asks. The aim of physics today is to give predictions concerning phenomena when we know all the effective factors (in mechanics, the forces) and initial conditions. This opens possibilities for ruling nature: by changing the effective factors and initial conditions, we can try to produce a given, desired situation. Instead of this, we could ask: How can it be understood that evolution has produced the situation which really is the case? We can try to *understand nature* instead of aiming to rule it.

In the investigation of evolution, for example, one should put more emphasis on the so-called *anthropic principle*, which states that the initial conditions of the universe (according to the Big Bang theories) seem to be precisely those needed to make life (in particular, human life) possible in the form we know life on earth. If the values of the fundamental constants of physics are changed even slightly, the existence of life in this form becomes impossible. It is as if the world had been 'made' in such a way that human beings could live in this universe. Physicists in general just shrug their shoulders at this fact, stating that "otherwise we would not be here." If we try to think of teleological problems, this fact is interesting, and perhaps such kinds of questions could open up new perspectives for science, helping us to understand our role in nature, instead of only wanting to control it.

Choice Between Two Beliefs

In fact, we have discussed here two possibilities, and the choice between them cannot be made on scientific grounds; the choice is by nature a decision of faith. Cosmologies are in general developed in a way which presupposes a belief that everything can be described with the aid of rational theories. Thus, one presupposes that laws of nature and 'chance' alone determine how evolution takes place.

Another possibility is a *belief* that there is purposefulness inherent in evolution, that it includes 'teleological guidance' which cannot be described by rational theories. Quantum physics opens this possibility to a purely scientific picture of the world. Although we build our world picture on a fully scientific (rational) basis, we can simultaneously believe that there is teleological guidance inherent in everything. We then look at evolution as a *continuous creation*: behind the rational aspects of evolution there is something irrational which determines the direction of evolution – the free choices in individual events.

Thus, one has reason to point out that the belief in creation has to be considered a matter that supplements the scientific conception of the world – if we have appropriated the lesson which atomic research has given us. The 'belief in chance' which still takes the place of teleology for most scientists can in no way be found to be more scientific. It is one expression of the one-sided rationalism.

As long as natural science was based on deterministic laws of nature, it was possible to justify the abandonment of every kind of teleology as 'non-scientific'. Statistical causality has changed the situation, and from now on whether we build our conception of the world on the 'religion of chance' or on the 'belief in creation' is purely a matter of faith. If our conscience forces us – when considering the difficult problems of human life – to choose the belief in creation, it becomes necessary for us to take a more critical stand towards the evolution theories, both in physics and in biology, than natural scientists in general seem to take.

For these kinds of reasons, I feel that it is irresponsible to use the authority of science for advertising a cosmology which is based explicitly on the 'religion of chance'. Science does not require it. Physicists (this author among them) are in general inclined to neglect the irrational dimensions of existence and to find an escape in a world view based on scientism. It can be a deceitful basis for life, and one should not recommend it to people who are searching for truth in quite another sense.

Cosmology is inseparably associated with a decision of faith. It is wrong if one tries to hide this behind a scientific form and if science is used as a means in propaganda which misleads people in such a choice – in the decision of faith – which decisively influences the direction of their life and the values which form the basis for the most important decisions they have to make. Scientists who one-sidedly adore reason are not the best possible guides in such questions.

4. The Power of Materialism

Petrified Attitudes

Finnish cultural life is still in materialistic shackles. Structures which originated in Finland at a time when a world view based on materialism was especially respected for political reasons still determine the basis for appraisal in many questions – without people even realizing it. I have always come up against this fact when I have tried to initiate discussion about the philosophical problems of atomic physics. Now is the time for criticism.

The atomic world cannot be understood without including in the picture of the world a fundamental element of quite another nature: consciousness. The nature of atomic phenomena seems to prove, indisputably, that materialism is a Utopia on a shaky foundation. 'Matter', when it is investigated more closely, is dissolved into abstractions which have very little to do with the conceptions of 'materiality' which are usually associated with materialism. The change concerning natural laws is particularly deep. When I tried to describe it in my book *Fysiikka ja usko* (Physics and Belief, *Laurikainen* 1978), I found that I had entered a *region which is absolutely forbidden* – or using the expression of Gregory Bateson, which „angels fear" *(Bateson* 1987).

Thereafter, the question 'Why?' has often occupied my mind. The criticism of materialism, which is the most important result of my scientific work, and also important from the point of view of social attitudes and values, is rejected on the basis of superficial propaganda. Must Finland still be a fortress of materialism where free discussion of the basic questions of knowledge and belief is not allowed? Does not society drift with the wind if these questions are solved through politics?

Western culture has broken reality into pieces, each of which has its guardians, and these have the right to decide what is true within that territory – but within that territory only. Nobody takes care of the whole. The Church originally had this obligation, but now theologians have fortified themselves in a territory of their own: the 'world of spirit'. Philosophy also has its own territory among the special sciences. Therefore, a very superficial materialism dominates in the natural sciences, where people just learn to speak jokingly of 'beings which have no body'.

It is time to ask, forcefully, why a free discussion is not wanted about questions which can influence the direction of science and its relation to religion, simultaneously clarifying the basis of values. *One should break up structures which bind us to a Utopia which does not have realistic grounds.*

The Wisdom of the Atoms

The power of materialism does not, of course, concern only the cultural milieu in Finland. In the natural sciences, materialism is considered the only possible pattern of thought. This is the explanation for why the profound ontological questions raised by quantum mechanics have remained unclear although they have been discussed for seven decades.

Recently, interest in the Copenhagen interpretation of quantum mechanics has again been increasing – very much in a negative sense, it is true. More and more physicists have found this interpretation in a strange way incomprehensible. On the other hand, so-called Einstein-Podolsky-Rosen (EPR) experiments have made it ever clearer that quantum mechanics presupposes a deep change in the conception of reality. Thus, enthusiastic research into the foundations of quantum mechanics is going on, in order to rescue the 'realistic' conception of reality. The strong development of the technology based on quantum physics in recent years now gives additional strength to these efforts. One even speaks of empirical philosophy.

The question now concerns precisely the possibility of materialism in atomic physics. It is not realistic to try to clear up the problems of atomic theory on a 'purely physical', i.e., on a materialistic, basis. This is now the heart of the discussion.

Almost all physicists are ready to abandon determinism, but in general they ignore the philosophical problems concerning individual events stating that even statistical laws make it possible to govern atomic phenomena with sufficient accuracy.

However, the result that individual events cannot be completely governed, even in principle, upsets the basis of the picture of the world. This implies that the 'truth' of a theory cannot mean that it, more or less exactly, describes 'objective reality'. In a scientific sense we can describe only our knowledge of reality: We can give statistical predictions concerning observable (macrophysical) phenomena which in some way refer to the atomic world, but that world itself we cannot reach.

This is a result upsetting the basis of realism. Reality remains forever veiled from human knowledge, but we can, however, form a reliable picture of a gross structure of reality on the basis of our experience, describing only *what we know* of reality. Thus, we can postulate a 'veiled reality', but we cannot compare our theory with it.

Knowledge presupposes *somebody who knows.* Therefore, atomic physics has given us a lesson that besides the 'matter' which the natural sciences investigate, there is a quite different element in reality: *consciousness*. It is, as human consciousness, inherent in all observations, in particular in their interpretations. But this different element of reality also appears in another, more general sense. Namely, all occurrences of the material world are, according to quantum physics, quantum events characterized by genuinely probabilistic laws. In individual material events, free choices between different possibilities always take place. In these 'choices' an aspect of reality appears which is absolutely rejected in the deterministic picture of the world. One should not compare the universe with a clock device – as has been done since the 17th century – but with an organism which has the ability to adapt and change. This is the basic lesson which the atoms have given us. It pulls out the foundation from under materialism.

Why Materialism Still?

My younger colleagues have in general answered such remarks just by a supercilious smile or by a remark that I am again trying to mix science with religion. When I have tried to propose that some courses in philosophy are needed in physics in order to provide a basis for problems of this kind as well, this has been absolutely rejected because there is neither time nor funds for teaching philosophy to physicists.

Since the 1960s, consciousness has always been excluded from discussion in physics because it is not possible to define this concept with the exactness which is required in physics. It is astonishing that in almost all publications concerning the foundations of quantum physics, consciousness is excluded from the discussion of observations. (Some exceptions to this rule have recently appeared, e.g., *Atmanspacher et al.* 1995, *d'Espagnat* 1983, *Laurikainen* 1993, *Primas* 1992, 1995, and *Stapp* 1993.) The exactness and objectivity of physics requires that one must reject concepts which cannot be defined and measured in the same way as normal physical quantities. A physicist must leave such matters to philosophers, people say.

Among the Finnish philosophers, only Eino Kaila (1890–1958) has really tried to understand the problems of quantum physics (e.g., *Kaila* 1942, 1950). Nowadays one obviously wishes to have a clear borderline between philosophy and physics – and then the basic problems of atomic physics do not really belong to anybody.

This is a rather international problem. One writes much about the foundations of quantum theory but problems concerning consciousness are avoided because it is difficult to describe them 'exactly enough' (not to mention the unconscious which for Pauli was the essence of the problem). One may well wait long for the exact definitions because the irrational cannot be made rational – and thus one is not allowed to speak of it!!

The problem is that *Cartesian dualism is not possible in atomic theory*. The spiritual element is inseparable from the description of the world of atoms. If we are realists in the true sense of the word, it is necessary to speak of it although we cannot describe it using a mathematical theory. There is an irrational element in reality which cannot be reached by the methods of the present natural science. If we acknowledge its existence we can ask many questions in a new way. The relation between man and nature is seen in a new light. Then we begin to understand why the present scientific–technical culture necessarily leads to a conflict with nature: It is based on a one-sided conception of reality where part of reality has been cut off – only because it cannot be described with mathematical accuracy. Therefore materialism dominates in physics and thus one claims that physics is value-free and objective.

What does this mean from the point of view of the relation between science and religion? It would need profound consideration, but this presupposes readiness for discussion from the side of both scientists and theologians. If theologians wish to dictate the truth in all questions concerning spirit, then an open discussion is not possible, of course. In many countries, the burning nature of these questions has been realized among both some scientists and some theologians. I hope that this will be realized in Finland too.

The essence of the problem is the power of materialism in science.

II

Facts and Interpretations

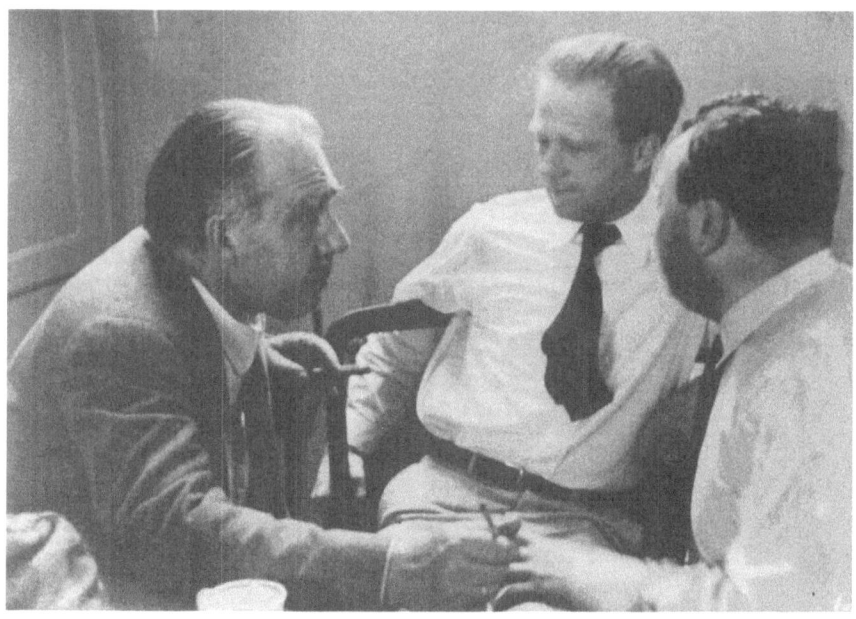

The kernel of the Copenhagen school: *Niels Bohr, Werner Heisenberg,* and *Wolfgang Pauli. Max Born* (Göttingen) was a close collaborator.
(By permission of the Niels Bohr Archive, Copenhagen.)

5. Ontology Implied by the Copenhagen Interpretation

In this chapter we wish to show, with the aid of quotations from original publications, that the philosophy of the Copenhagen interpretation is usually misunderstood today. This is partly due to the different views of Bohr and Pauli concerning ontology and the role of the observer. Pauli's philosophy was more consistent. Pauli called the psychophysical problem the most important question of our time. If we are not bound to materialistic presuppositions, the psychophysical *unus mundus* of Pauli and Jung appears, in fact, to be the natural ontology of quantum mechanics.

The main result in this chapter is that it is wrong to claim that physics describes independent reality. Although the aim of physics is to describe objective reality, the observation process and its interpretation presuppose the activity of the human consciousness. Therefore our description concerns only *our knowledge* of reality.

Another important point emphasized by Bohr and Pauli is the *wholeness* (Bohr: individuality) of a quantum process. Every attempt to subdivide a quantum process – to investigate its details – presupposes a change in the necessary experimental arrangements and, a fortiori, a change of the state function. This means that we are no longer studying the original process. (See especially *Pauli* 1984, p. 58 below.)

Is the Observer Detached?

It is not generally known that there was a profound difference in the philosophical attitudes of Niels Bohr and Wolfgang Pauli (*Laurikainen* 1985b, Sect. 3). In his address at the Second Centenary of Columbia University in 1954, "The Unity of Knowledge", Bohr claimed that the observer even in quantum mechanics can be considered 'detached' provided we understand the observation in the right way (*Bohr* 1955, p. 83). An observation includes a detailed description of all the experimental arrangements which can have an influence upon the phenomenon under investigation, and it is finished only when a registered result is obtained which everybody can verify afterwards. In this sense, Bohr said, an observation is quite *objective* (which for Bohr means the same as 'intersubjective'), and the observer does not have any influence on the result in any other way than by choosing the method of observation. The result is explicitly associated with a given method of observation. If physics is understood as a system which makes it possible to govern such objective observational results – which, however, is only possible in a probabilistic sense – then physics, according to Bohr, can even in atomic physics be considered quite objective and the observer is 'detached' in exactly the same way as in classical physics.

Pauli had a different opinion. In his letter to Bohr of February 15, 1955, where he made remarks about Bohr's article concerning the Columbia address, Pauli pointed out that the role of the observer in quantum mechanics is essentially different to that in classical physics because an observation changes the 'state' of the observed system in a way which cannot be predicted unambiguously by using any laws of nature *(Laurikainen* 1985b, p. 282; *Laurikainen* 1988, p. 60):

I think here on the passage to a new phenomenon by observation which is technically taken into account by the so-called "reduction of the wave packets". As it is allowed to consider the instruments of observation as a kind of prolongation of the sense organs of the observer, I consider the impredictable change of the state by a single observation – in spite of the objective character of the result of every observation and notwithstanding the statistical laws for the frequencies of repeated observation under equal conditions – to be *an abandonment of the idea of the isolation (detachment) of the observer from the course of physical events outside himself.*

To put it in nontechnical common language one can compare the role of the observer in quantum theory with that of a person, who by its freely chosen experimental arrangements and recordings brings forth a considerable "trouble" in nature, without being able to influence its unpredictable outcome and results which afterwards can be objectively checked by everyone.

Pauli here emphasizes the fact that the observer interferes (in quantum mechanics) essentially in the process of observation and thus he cannot be considered 'detached' as in classical physics. In spite of Pauli's remarks Bohr maintained in his article the claim that the observer, even in quantum mechanics, is 'detached' in the same sense as in classical physics. This means that Bohr accepts the idea of independent reality which the observer examines with the aid of his observations. He stated, of course, that it is not possible to describe the properties of this 'reality' in a more detailed way than in quantum mechanics, and this forced him to avoid ontological considerations and speaking of realism in the traditional sense of the word.

Thus, Bohr confined himself to epistemology, consciously avoiding ontology. Abner Shimony described in Helsinki in 1992 the philosophical attitude of Bohr very nicely in exactly this sense (*Shimony* 1993, pp. 88–94 & 175–187).

Bohr described observation as an interaction between the observed object and the measuring equipment, and he especially warned of mixing psychological considerations with the interpretation of a physical theory because otherwise we should lose the *objectivity* of physics, which Bohr especially emphasized.

Pauli could not understand this change which seemed to have taken place in Bohr's attitude. When describing complementarity Bohr used to emphasize that when making an observation we essentially influence the process which we can observe, stating that "in the great drama of existence, we are not just spectators but also actors". According to Pauli we cannot neglect the psychic activity of the observer because the observer can freely choose the method of observation and the observer also interprets the results. In his letter to Fierz of October 13, 1951 Pauli describes the inseparable role of the observer as follows (*Laurikainen* 1988, p. 35):

Now there comes the major crisis of the quantum of action: One has to sacrifice the unique individual and the "sense" of it in order to save an objective and rational description of the phenomena. If two observers do the same thing even physically it is, indeed, really no longer the same: Only the *statistical averages* remain, in general, the same. *The physically unique individual is no longer separable from the observer* – and for this reason it goes through the meshes of the net of physics. The individual case is *occasio* and not *causa*. I am inclined to see in this *occasio*, which includes within itself the observer and the selection of the experimental procedure which he has hit upon, a revenue of the *anima mundi* which was pushed aside in the seventeenth century (naturally "in an altered form"). La donna è mobile – so are the *anima mundi* and the *occasio*.

Here Pauli wishes to point out that the observer, by planning the method and by his decisions, influences the results in an unpredictable way: The observer's psychic functioning is mixed with the phenomenon. In several connections Pauli stated that quantum phenomena force us to abandon the *psychophysical parallelism* which he called a 'spiritual cloud of fog' in Western thought (*Laurikainen* 1988, p. 141).

Thus, Pauli could not accept Bohr's claim that the observer is 'detached' in the process of observation (*Pauli* 1984, pp. 7, 8). Besides the fact that the observer essentially interferes with the phenomenon which he observes, Pauli always presupposes that the interpretation also belongs to the observation. *The reduction of the state function does not take place when the detector reacts to the situation (for example, certain signs are formed on a magnetic tape) but only when the result has been interpreted and the observer has become conscious of it.* (See also *Heisenberg* 1959, p. 38 and *Wigner* 1967, p. 172.)

According to Pauli an observation is not just a physical interaction between the object and the measuring apparatus but some kind of interaction between the object and the consciousness of the observer. This 'interaction' cannot be described by the means of the now-prevailing science because it is an 'interaction' between a material object and the psyche of the observer. In the interpretation of physics is included here something which physicists are not accustomed to think of: the *consciousness*. As far as I understand, this is a necessary presupposition of the Copenhagen interpretation. In addition to Pauli, at least Heisenberg and Born have emphasized that quantum mechanics does not describe the atomic world but is a description of *what we know* of the atomic world (*Heisenberg* 1955, pp. 12, 18; *Born* 1961, pp. 457, 458; *Born* 1983, p. 298). This presupposes that, besides the atomic world, one speaks also of consciousness and, thus, of the psyche of the observer.

Here, in the 1950s, Bohr took a cautious attitude because he feared – with good reason – that the inclusion of the psyche would mean the introduction of a certain subjectivity into physics. One can ask, however, whether the objectivity of physics can be rescued by just avoiding discussion of psychic matters although it is quite obvious that such a discussion is necessary. It is possible that Bohr's caution was partly due to the fact that the Copenhagen interpretation was strongly criticized in the Soviet Union at the beginning of the 1950s and this had occasioned difficulties for some physicists.

The Psychophysical Reality

At least Heisenberg has, like Pauli, pointed out that quantum mechanics requires the abandonment of Cartesian dualism (*Heisenberg* 1959, p. 66). Observations do not reveal the objective reality but a picture of reality embellished by our psyche. This is a fact which Kant already emphasized, with the difference, however, that in physics we now believe that observations give information about something which must be called 'real'. Kant claimed that empirical knowledge cannot at all reach 'reality itself' but just the *phenomenal world* created by the human psyche. In quantum mechanics we can say that we obtain certain knowledge about 'reality itself' – as far as it is meaningful to speak of 'reality' – but from the point of view of empirical science this reality remains as behind a veil: We cannot describe reality itself but only what we know of reality. An expression of this situation is that natural laws are not deterministic but statistical. We can present probabilistic expectations for real phenomena and even judge the reliability of these expectations.

This new situation arises from the necessity to introduce the psychic element into the conception of reality. The world that we describe in empirical science is a world shaped by the functioning of the human psyche, quite like the phenomenal world of Kant, because observations are results of psychic activity. D'Espagnat has called the picture of the world given by science *empirical reality* (*d'Espagnat* 1993, p. 56). It is a human picture of the 'reality itself' which science can never reach.

The reality itself d'Espagnat calls *independent reality*. Scientists generally believe that there is a reality independent of any observations. If we believe that an observer gets quite objective results, as the idea of the detached observer presupposes, then we must assume that such an independent reality exists. Einstein writes that this is like an axiom for natural sciences: It cannot be proved but it simply is accepted as a starting point of research (*Fine* 1986, p. 95). In this sense Bohr, Born, Heisenberg, and Pauli all *believed* in the existence of an independent reality. D'Espagnat also emphasizes the importance of independent reality in his article "Open Realism" (*d'Espagnat* 1993, p. 52). Characteristic of all of them is the attitude corresponding to open realism: One may not presuppose anything more about reality than what can be proved experimentally. The most cautious one was Bohr who did not like to speak at all of ontological problems. Actually his attitude presupposed, however, an ontological hypothesis: He presupposed that it is possible to describe physical phenomena without taking psychic matters into account. This hypothesis is based on Cartesian dualism and has strengthened the materialistic attitude of physicists. As Bohr simultaneously emphasized complementary aspects of reality, it is difficult to understand how one can neglect the influence of psyche on observations.

Since natural laws have been found *genuinely statistical*, individual events cannot be governed theoretically. Therefore Pauli states that independent reality contains irrational elements: It is partly rational, as can be seen from the existence of natural laws, but it also contains irrationality which finds its expression in the 'freedom' or 'choices' characteristic of the individual events. Rationality and irrationality, together, are one expression of the complementarity of reality, and the idea of statistical causality contains both of these aspects of reality (*Pauli* 1984, pp. 21–22).

On the basis of all this, it is not difficult to understand that Pauli considered reality to be *psychophysical* (*Pauli & Jung* 1992). Empirical reality is, of course, psychophysical because it is shaped by the human psyche. However, because natural phenomena obey statistical laws, this leads us to think that independent reality or 'reality itself' must also have both rational and irrational aspects – resembling the activity of the human psyche (cf. *Laurikainen* 1993 and *Pauli & Jung* 1992).

The Importance of the Conception of Reality for Science and for Culture in General

In this way, the properties of independent reality begin to acquire features characteristic of a living organism: One can trace expressions of *reason* and *will* – or, according to Schopenhauer's well-known book, *Wille und Vorstellung* ('will' and 'idea'). In addition, *teleology* becomes possible in phenomena, a fact which corresponds to the behavior of living organisms and to their ability to adapt themselves to the circumstances.

Ontological remarks of this kind have the property of directing our thought. The deterministic conception of causality in the 17th century produced the idea of a mechanical world which has been characteristic of scientific thought up to the present. Although physics has abandoned the idea of deterministic causality, the changes in research, which this should imply, have not at all taken place yet. I have repeatedly pointed out that people continuously develop cosmological models on the basis of the Einsteinian belief that reality can be described with the aid of mathematical theories, even from the very 'beginning of time'.

If one thinks that reality is analogous to an organism, the evolutionary theories must be developed much more carefully. One must find quite new, teleological principles which are not aimed at manipulating events but at understanding the nature of reality. The cosmological anthropic principle is a good beginning. The present theories of evolution contain implicitly dogmatic (materialistic) elements; especially in the popularization of these theories, one should clearly point these out.

Research in the natural sciences generally presupposes traditional realism in some form, i.e., one tries to describe independent reality. In addition, Cartesian dualism has a strong influence on the direction of research in the sense that psychic matters are quite generally neglected, at least when investigating inanimate nature. People have so far totally neglected psychic matters in the basic research of quantum mechanics. This is a requirement of materialistic philosophy, and even Bohr accepted this attitude in the 1950s. On the basis of the remarks made above, this is quite impossible to understand. This is a good example of how the (usually implicit) conception of reality strongly influences the direction of research. There is reason to think that Pauli's conception of psychophysical reality can open quite new perspectives for the basic research into quantum mechanics and particle physics, for example. These fields, as also the great question of combining the principles of quantum physics and of the theory of relativity, are related to our views on space, time, and causality and, a fortiori, to the conception of reality.

I would like to point out that the EPR correlations, which have been extensively discussed, can simply be understood as results of *shaping procedures* if we accept the psychophysical conception of reality. More generally, 'matter waves' or the quantum mechanical state functions can be interpreted as symbolic means in the shaping activity of the psyche. The notorious reduction of the state function does not take place in the outer world but *in our consciousness.* If one, instead, imagines 'matter waves' as waves proceeding in space and time, more or less analogously to mechanical waves, one is easily misled as to questions of superluminal (faster than light) velocities, etc. The materialistic conception of reality has indeed led research astray here.

Bohr's attitude, when he tried to avoid ontological problems and to confine himself to epistemological questions only, can be criticized for the very reason that it has implicitly influenced the direction of research. The result has been that the basic research of quantum mechanics has been led to unreal problems (I have occasionally called them modern 'theories of epicycles'). For this reason I also find it very questionable that one today puts so much weight on the philosophy of language. Even in this way one avoids the discussion of ontological questions, and very easily this leads to diminishing the importance of psychic problems and, a fortiori, to a path dominated by materialism.

We cannot here go on to the description of the psychophysical reality. As is known, Pauli became interested in Jung's depth psychology and especially in the concept of archetype (*Pauli* 1954; *Pauli & Jung* 1992). According to Pauli, archetypes are basic elements of the psychophysical reality, unus mundus, in the same sense as ideas are basic elements in Plato's ontology. The important difference is that archetypes have an essentially dynamic nature: They create new shapes (Gestalt) and themselves undergo changes, while Plato's ideas are invariant. *Unus mundus is an evolutionary world.*

The ontology formed on the basis of quantum mechanics thus leads to a *complementary, psychophysical conception of reality* in place of Cartesian dualism. This change in the conception of reality may have a revolutionary influence on Western thought. One important implication is a more positive relation between science and religion. Both of them can be understood as different (complementary) ways of approaching one and the same reality.

This chapter was published in Finnish in Arkhimedes 2/1994 and in English in the Appendix of *Symposium* 1994.

Original Texts of the Quotations Referred to in Chap. 5

The quotations are given in their original form, without any corrections being made. They are given in the same order as they appear in the main text. Quotations given in the main text are not included.

It is important that the reader studies carefully the wordings in these quotations and compares them with the main text in this chapter because then it should be clear that the philosophy of the Copenhagen interpretation is quite generally misunderstood today. All of the creators of this interpretation agree that the quantum mechanical description does not describe the atomic world but only our knowledge of this world; Bohr calls this description 'symbolic'. The differences between the main philosophies of the founding fathers, which are strongly emphasized by some authors, appear to be unessential if these quotations are studied carefully. The difference between Bohr and Pauli concerning the role of the observer must be pointed out, however.

Laurikainen 1985b (Proceedings, *Symposium 1985*, p. 279.)
This quotation is paraphrased in the main text.

Bohr 1955 (*Atomphysik und menschliche Erkenntnis*, 1985, p. 83.)
Der Begriff Komplementarität bedeutet in keiner Weise ein Verlassen unserer Stellung als aussenstehende Beobachter, er muss vielmehr als logischer Ausdruck für unsere Situation bezüglich objektiver Beschreibung in diesem Erfahrungsbereich angesehen werden. Die Erkenntnis, dass die Wechselwirkung zwischen den Messgeräten und den untersuchten physikalischen Systemen einen integrierenden Bestandteil der Quantenphänomene bildet, hat nicht nur eine unvermutete Begrenzung der mechanistischen Naturauffassung, welche den physikalischen Objekten selbst bestimmte Eigenschaften zuschreibt, enthüllt, sondern hat uns gezwungen, bei der Ordnung der Erfahrungen dem Beobachtungsproblem besondere Aufmerksamkeit zu widmen.

Shimony 1993 (Proceedings, *Symposia 1992*, pp.79–96 and 175–191.)
This quotation is too long to be presented here.

Laurikainen 1988, p. 141. (Pauli an Fierz, 3.6.1952. PLC 0092.092)
PLC refers to the Pauli Letter Collection at CERN (see the list of literature at the end of the book). *Laurikainen* 1988 gives several quotations from unpublished letters in PLC (identified according to the bookkeeping of the Collection).
Das hier "konstellierte" zentrale Problem ist m.E. das "*psycho–physische*". Mehr und mehr kam ich zur Überzeugung, dass der im Anschluss an Leibniz u. Spinoza ausgebildete Begriff des "Parallelismus", vom Standpunkt der *klassischen Physik* aus betrachtet, illegitim und "erschlichen" wird ... Denn wenn alles deterministisch-kausal sein soll, gibt es m.E. keinen Platz für eine andere Art von Zusammenhang, die etwa statt mit "kausal" mit "parallelistisch" zu bezeichnen wäre. Daher das Vorhandensein des "psycho-physischen Parallelismus" getauften geistigen Nebelfleckes ebenso ein Hinweis auf die

Unvollständigkeit des klassisch-naturwissenschaftlichen Weltbildes ist wie z.B. der licht-elektrische Effekt und das Wirkungsquantum. Es ist mir daher befriedigender zu denken, dass es die *akausale* Art des Zusammenhanges, die "psycho-physischer Parallelismus" genannt wurde, qua "Angeordnet-sein" bezw. "Korrespondenz" **sonst auch** geben muss und nicht nur speziell bei Psyche–Physis.

Pauli 1984 (*Physik und Erkenntnistheorie*, pp. 7–8.)
Die Bedeutung dieser Entwicklung besteht darin, dass sie uns einen Einblick in die logische Möglichkeit einer neuen und erweiterten Art des Denkens gibt. Dieses Denken zieht auch den Beobachter mit in Betracht, einschliesslich des von ihm benutzten Messgerätes, also ganz verschieden von der Art, wie es in der klassischen Physik geschah, sowohl in der *Newton*schen Mechanik als auch in den *Maxwell–Einstein*schen Feldtheorien. Bei der neuen Art, zu denken, nehmen wir nicht mehr den *losgelösten Beobachter* an, der in den Idealisierungen dieser klassischen Theorietypen auftritt, sondern einen Beobachter, der durch seine undeterminierbaren Einwirkungen eine neue Situation schafft, die theoretisch als ein neuer Zustand des beobachteten Systems beschrieben wird. Auf diese Weise ist jede Beobachtung eine Aussonderung eines realen Einzelereignisses, hier und jetzt, aus den theoretischen Möglichkeiten, wobei gleichzeitig die diskontinuierliche Seite der physikalischen Phänomene zu Tage tritt.

Indessen bleibt auch in der neuen Art von Theorie immer noch die *objektive Wirklichkeit*, in sofern als diese Theorien jede Möglichkeit leugnen, dass der Beobachter das Ergebnis eines Experiments noch beeinflussen kann, nachdem einmal die experimentelle Anordnung gewählt worden ist. Deshalb gehen die besonderen Eigenschaften eines individuellen Beobachters nicht in den begrifflichen Rahmen der Theorie ein. Im Gegenteil beschreibt sie die Erscheinungen in der mikroskopischen Grössenskala atomarer Objekte auf eine Weise, die für jeden beliebigen Beobachter gilt, und mit mathematischen Gesetzen mit gruppentheoretischen Eigenschaften, die ein jeder erlernen kann, der über ausreichende mathematische und physikalische Kenntnisse verfügt. In diesem erweiterten Sinne ist die quantenmechanische Beschreibung atomarer Phänomene immer noch eine objektive Beschreibung, obgleich nicht mehr angenommen wird, dass der Zustand eines Objekts unabhängig bleibt von dem Weg, auf dem die möglichen Quellen einer Information über das Objekt durch Beobachtungen unwiderruflich geändert werden. Die Existenz solcher Änderungen enthüllt uns eine neue Art von Ganzheit in der Natur, die der klassischen Physik unbekannt war, insofern als ein Versuch, das durch die ganze zur Beob-achtung verwendete experimentelle Anordnung definierte Phänomenen zu unterteilen, ein ganz und gar neues Phänomen schafft. [Italics is used instead of Pauli's spaced-out text.]

Heisenberg 1959, p. 38. (*Gesammelte Werke*, C II, p. 38.)
Der Übergang vom Möglichen zum Faktischen findet also während des Beobachtungsaktes statt. Wenn wir beschreiben wollen, was in einem Atomvorgang geschieht, so müssen wir davon ausgehen, dass das Wort 'geschieht' sich nur auf die Beobachtung beziehen kann, nicht auf die Situation zwischen zwei Beobachtungen. Es bezeichnet dabei den physikali-schen, nicht den psychischen Akt der Beobachtung, und wir können sagen, dass der

Übergang vom Möglichen zum Faktischen stattfindet, sobald die Wechselwirkung des Gegenstandes mit der Messanordnung, und dadurch mit der übrigen Welt, ins Spiel gekommen ist. Der Übergang ist nicht verknüpft mit der Registrierung des Beobachtungsergebnisses im Geiste des Beobachters. Die unstetige Änderung der Wahrscheinlichkeitsfunktion findet allerdings statt durch den Akt der Registrierung; denn hier handelt es sich um die unstetige Änderung unserer Kenntnis im Moment der Registrierung, die durch die unstetige Änderung der Wahrscheinlichkeitsfunktion abgebildet wird.

Wigner 1967 (*Symmetries and Reflections*, p. 172; quoted from *Wigner* 1962.)
When the province of physical theory was extended to encompass microscopic phenomena, through the creation of quantum mechanics, the concept of consciousness came to the fore again: it was not possible to formulate the laws of quantum mechanics in a fully consistent way without reference to the consciousness.[3] All that quantum mechanics purports to provide are probability connections between subsequent impressions (also called "apperceptions") of the consciousness, and even though the dividing line between the observer, whose consciousness is being affected, and the observed physical object can be shifted towards the one or the other to a considerable degree,[4] it cannot be eliminated. It may be premature to believe that the present philosophy of quantum mechanics will remain a permanent feature of future physical theories; it will remain remarkable, in whatever way our future concepts may develop, that the very study of the external world led to the conclusion that the content of the consciousness is an ultimate reality.

[3] W. Heisenberg expressed this most poignantly [*Daedalus*, 87, 99 (1958)]: "The laws of nature which we formulate mathematically in quantum theory deal no longer with the particles themselves but with our knowledge of the elementary particles." And later: "The conception of objective reality ... evaporated into the ... mathematics that represents no longer the behavior of elementary particles but rather our knowledge of this behavior." The "our" in this sentence refers to the observer who plays a singular role in the epistemology of quantum mechanics. He will be referred to in the first person and statements made in the first person will always refer to the observer.

[4] J. von Neumann, *Mathematische Grundlagen der Quantenmechanik* ...

N.B. Notes belong to the quotation.

Heisenberg 1955 (*Das Weltbild der heutigen Physik*, pp. 12 & 18.)
p. 12. Dies hat schliesslich zur Folge, dass die Naturgesetze, die wir in der Quantentheorie mathematisch formulieren, nicht mehr von den Elementarteilchen an sich handeln, sondern von unserer Kenntnis der Elementarteilchen.

p. 18. ..., dass wir die Bausteine der Materie, die ursprünglich als die letzte objektive Realität gedacht waren, überhaupt nicht mehr "an sich" betrachten können, dass sie sich irgend einer objektiven Festlegung in Raum und Zeit entziehen und dass wir im Grunde immer nur unsere Kenntnis dieser Teilchen zum Gegenstand der Wissenschaft machen können.

Born 1961 (*Werner Heisenberg und die Physik unserer Zeit*, p. 457.)
Vielmehr ist die folgende Auffassung möglich. Die Naturgesetze beziehen sich auf unser

Wissen über die Objekte. Dieses Wissen ist jederzeit unvollständig und ungenau. Die sogenannten Naturgesetze erlauben, aus dem Wissen über den Zustand zu einer Zeit Aussagen über das zu einer andern Zeit zu Erwartende zu machen.

Dies wird allen praktisch tätigen Naturforschern (und Ingenieuren) selbstverständlich erscheinen, ist aber vom philosophischen Standpunkte ein Wechsel der Einstellung, dessen Radikalität mir nur langsam bewusst geworden ist.

Born 1983 (*Physik im Wandel meiner Zeit*, p. 298.)
Die hinter meiner Theorie liegende Philosophie habe ich noch Jahre lang durchdacht und dann ganz knapp in der Festschrift zu *Heisenbergs* 60. Geburtstag dargelegt. Sie läuft darauf hinaus, dass wissenschaftliche Vorhersagen sich gar nicht direkt auf die "Wirklichkeit" beziehen, sondern auf unser Wissen von der Wirklichkeit. D.h. die sogenannten "Naturgesetze" erlauben, aus dem augenblicklichen beschränkten, angenäherten Wissen auf eine zukünftige, natürlich auch nur approximativ beschreibbare Situation zu schliessen. Das ist eine Denkweise, die der *Einsteins* schroff entgegensteht, und es ist kein Wunder, dass er mich als Abtrünnigen ansah. Und doch habe ich das Gefühl, dass ich treu den Weg weitergegangen bin, den er uns in seiner grossen Zeit gewiesen hat, während er an einer bestimmten Stelle stehen geblieben ist. Diese Stelle ist die Vorstellung, dass die Aussenwelt, wie sie wirklich ist, von der Wissenschaft getreu und genau beschrieben wird. Von diesem Gesichtspunkt aus ist die heutige Theorie der Materie in der Tat ein Wirrsal von Absurditäten, und *Einstein* hatte von seinem Standpunkt ganz recht, sie abzulehnen oder höchstens als Provisorium gelten zu lassen.

Heisenberg 1959, p. 66. (*Gesammelte Werke*, C II, p. 66.)
Wenn man an die grossen Schwierigkeiten denkt, die selbst bedeutende Naturwissenschaftler wie Einstein bei dem Verständnis und der Anerkennung der Kopenhagener Deutung der Quantentheorie hatten, so kann man die Wurzeln dieser Schwierigkeit bis zur cartesianischen Spaltung verfolgen. Diese Spaltung ist in den drei Jahrhunderten, die auf Descartes gefolgt sind, sehr tief in das menschliche Denken eingedrungen, und es wird noch lange dauern, bis sie durch eine wirklich neue Auffassung vom Problem der Wirklichkeit verdrängt ist.

d'Espagnat 1993 (Proceedings, *Symposia 1992*, p. 56.)
Hence the object of scientifically fruitful descriptions of physics should be given another name. *Empirical* or *effective* reality, for example. And what emerges from all this is that both notions of (veiled) reality-*per-se* and empirical reality must be considered as significant.

Fine 1986 (Einstein's letter to Laserna, 8.1.1955, see Arthur Fine: *The Shaky Game: Einstein Realism and the Quantum Theory*, p. 95.)
It is basic for physics that one assumes a real world existing independently from any act of perception. But this we do not *know*. We take it only as a programme in our scientific endeavors. This programme is, of course, prescientific and our ordinary language is already based on it.

d'Espagnat 1993 (Proceedings, *Symposia 1992*, p. 52.)
For these, and a few other, reasons, I can accept neither neo-Kantianism nor any of the philosophical theories that similarly reject the notion of anything logically anterior to our human experience and somehow inducing it. Let me give a name to the standpoint I am let to this way. Let me call it "open realism" (B. d'Espagnat, *Dialectica*, **43** Fasc. 1–2 (1989) 157):

Open realism
Existence is logically prior to knowledge.
Some will agree with me that this is quite obvious. Others will call it a postulate. For me it does not matter very much. What, in my approach, I consider as essential is that I do not make any other postulate. I mean: contrary to many conventionalist realists I do not postulate anything concerning the nature of "what exists"; not even that it is knowable. Nor do I postulate that it is unknowable! I just leave the matter open: it is to be decided on the basis of what factual knowledge, and in particular scientific knowledge, will reveal.

Pauli 1984 (*Physik und Erkenntnistheorie*, pp. 21–22.)
Hierdurch bekommt die Beobachtung den Charakter der *irrationalen, einmaligen Aktualität* mit nicht vorhersagbarem Resultat. Überdies bedingt die Unmöglichkeit, die Versuchsanordnung zu unterteilen, ohne das Phänomen wesentlich zu ändern, einen neuen Zug von *Ganzheitlichkeit* im physikalischen Geschehen. Diesem *irrationalen* Aspekt der konkreten Erscheinungen, die der *Aktualität* nach festgestellt sind, steht gegenüber der *rationale Aspekt* einer abstrakten Ordnung der *Möglichkeiten* von Feststellungen mit Hilfe des mathematischen Wahrscheinlichkeitsbegriffes und der ψ-Funktion.

... Diese logische Verallgemeinerung hat sich unter dem Druck der unter dem Stichwort "Endlichkeit des Wirkungsquantums" zusammengefassten physikalischen Tatsachen als schliesslich befriedigende Lösung früherer Widersprüche in einer höheren Synthese herausgebildet: Die mathematische Erfassung der *Möglichkeiten* des Natur-geschehens in der Quantenmechanik erwies sich als ein genügend weiter Rahmen, um auch die irrationale *Aktualität* des Einmaligen aufzunehmen. Als Zusammenfassung des rationalen und des irrationalen Aspektes einer wesentlich paradoxen Wirklichkeit kann sie auch als eine Theorie des Werdens bezeichnet werden.

[Here italics is used instead of Pauli's spaced-out text.]

Pauli & Jung 1952 (Wolfgang Pauli und C.G. Jung: *Ein Briefwechsel 1932–1958*. The psychophysical reality is a central theme in this volume.)

Laurikainen 1993 (Proceedings, *Symposia 1992*, pp. 207–218.)

Pauli 1954 (*Physik und Erkenntnistheorie*, pp. 113–128.)

The last three quotations are too long to be presented here.

6. Basic Features of Wolfgang Pauli's Philosophy

There is no systematic presentation of Pauli's philosophical thought, and therefore one must form a view of it on the basis of Pauli's rather concise articles concerning epistemological and ontological questions and from his extensive correspondence. The review given here is based on work done during nearly 20 years of study in the area of Pauli's philosophy. I have given a more comprehensive description of the subject in my book *Beyond the Atom. The Philosophical Thought of Wolfgang Pauli* (*Laurikainen* 1988) and in the Conference Proceedings *Symposia on the Foundations of Modern Physics 1992: The Copenhagen Interpretation and Wolfgang Pauli* (*Symposia 1992*) and *Symposium on the Foundations of Modern Physics 1994: 70 Years of Matter Waves* (*Symposium 1994*).

Introduction

Pauli's philosophical views are based on atomic theory (quantum mechanics). He adopted wholeheartedly Niels Bohr's idea of complementarity and developed it further toward an ontology. In this respect, a difference in views appeared between Bohr and Pauli in the 1950s concerning the role of the observer in microphysics (see "Wolfgang Pauli and the Copenhagen Interpretation" by the present author in *Symposium* 1985). While Bohr avoided taking a stand with respect to the conception of reality and realism (*Shimony* 1992a,b), Pauli adopted a view of reality in which physical and psychic phenomena are understood as complementary expressions of an abstract, non-describable reality. *Reality itself is considered transcendent* (incomprehensible, unattainable to rational theories). In addition to the rational features which science describes, reality contains irrationality which cannot be comprehended by scientific methods (*Symposia 1992*, Appendix and *Laurikainen* 1988, index entry for "irrational").

Characteristic of Pauli's thought is that he found the basic direction of Western culture to need a correction because "they went a little too far in the 17th century" (*Laurikainen* 1988, p. 40). The aim of this chapter is to show that Pauli's philosophy does not fit into the frame of the concept of science which is generally accepted in the scientific world today. This has caused misunderstandings and repression of the natural philosophy based on Pauli's thought. One hopes it will be possible to overcome the restrictions which a materialistic conception of reality and an exaggerated rationalism have so far caused in the opinions of the scientific community.

The Irrationality of Reality

The idea of the irrationality of reality is central from the point of view of Pauli's philosophy. This idea seems to elicit criticism based simply on passion rather than on thoughtfulness, which is not unexpected, of course, because Pauli in this respect deviates from the mainstream of Western philosophy which, since the Eleatics and Plato, has considered rationality (comprehensibility) a characteristic property of reality.

As justification for the irrationality of reality, Pauli presents the probabilistic laws of atomic physics, which show that the deterministic causality characteristic of classical physics has to be replaced on the microlevel by a more general conception of causality (*Laurikainen* 1988, pp. 32–35). Bohr often stated this fundamental idea by saying that causality is not valid in the microworld but must be generalized into the 'framework of complementarity' (*Bohr* 1985, e.g., pp. 17, 18 and article 8). Pauli spoke of the more general concept of statistical causality, which replaces classical, deterministic causality and contains the classical form of causality as the special case where 'probabilities are like one'. This means that laws do not refer to individual events but to large (in principle infinitely large) statistics. Instead of complementarity, Pauli preferred to emphasize this new concept of causality as the most characteristic feature of atomic theory (see editorial by Pauli in the special issue dedicated to complementarity in *Dialectica* (1948), Vol. 2, No 3/4).

The idea of statistical causality means that conformity to a law concerning statistics (a very large ensemble) is considered a 'last fact' without any explanation, corresponding to the very nature of reality (*Laurikainen* 1988, p. 32). Pauli often emphasized that quantum mechanical probabilities are *primary* ones, in contrast to the probabilities of classical statistical mechanics where probabilities are only introduced because the detailed control of large ensembles exceeds the possibilities of human knowledge. The Copenhagen interpretation has from the very beginning presupposed that the appearance of probabilistic laws in microphysics is not an expression of an incomplete knowledge but corresponds to the very nature of real processes (*Pauli* 1994, e.g., p. 32). Einstein, as is well known, held a different opinion, and the famous dispute between Bohr and Einstein about the 'completeness' of quantum mechanics concerned exactly this primary nature of the probabilistic laws (see *Shimony* 1992a).

The concept of statistical causality thus means that the detailed description of individual events is not possible, even in principle. They contain a freedom of principle insofar as each microsystem seems to be able to 'freely choose' one of the possible states which are allowed according to the probabilistic law. This 'free choice' in quantum mechanical formalism is represented by the famous *reduction of state.* It is a typical expression of the irrationality of reality: This choice does not have any rational explanation. (*Pauli* 1994, e.g., p. 32. Cf. also *Laurikainen* 1988, pp. 32–35.) Einstein, and in his footsteps most people working in the field of the foundations of quantum theory since the 1960s (after the death of Bohr), have not accepted this irrationality but instead have tried to describe the microworld so that the reduction of state receives some rational explanation. Not one of the new approaches has succeeded in this endeavor, however (see, e.g., *Shimony* 1992b).

The idea of statistical causality is not confined to microphysics, of course. There is even more reason to be critical with respect to deterministic laws in the field of life phenomena and especially in the humanistic or social sciences. The 'freedom' characteristic of individual events is especially emphasized within life phenomena because living organisms are able to adapt to circumstances, and the possibility of making choices is very clearly seen in life phenomena.

According to Pauli, the 'freedom' characteristic of individual events is the most important teaching of quantum mechanics. He often referred to the philosophy of Schopenhauer where the basic elements of reality are will and idea (Wille und Vorstellung), i.e., the (irrational) freedom of choice and the (rational) idea (*Laurikainen* 1988, e.g., pp. 32, 102 and 145–147). The latter represents conservation (invariance), the former represents changes. Although changes (phenomena) have been the most important objects of research in the modern ages (since Galileo's time), at the center of interest have been the invariant laws governing phenomena. According to Pauli, "they went a little too far in the seventeenth century", and the result was a deterministic world machine. Atomic research forces us to realize that there is in phenomena a feature resembling free will, and this is the origin of real changes in the world. Pauli has emphasized that this once more brings teleology into sciences. The general belief in random choices as the only possibility is not scientifically motivated. (*Laurikainen* 1988, p. 136 and Chs. V–VII. *Pauli* 1994, articles 16 and 17.)

It is clear that here the border line between knowledge and belief becomes obscure, and this Pauli found quite natural. He wrote that the sources of science and religion are the same (*Laurikainen* 1988, p. 135). Both of them have as their aim the comprehension of reality. *Rational science and mystical experience represent two complementary ways of understanding reality.* Pauli emphasized that it is impossible to draw a strict boundary between physics and metaphysics and that such an attempt leads to a dangerous narrowness:

Man soll sich immer bewusst bleiben, dass es für alles Vernünftige Geschmacksache ist, was man 'Metaphysik' und was man sonst wie nennt. (From Pauli's letter to Fierz on 11.4.1953. PLC 0092.109.)

The Complementarity of Reality

Quantum mechanics is throughout a complementary theory. The starting point for the idea of complementarity is particle–wave dualism, which means the use of two mutually exclusive pictures in the description of microparticles: a particle, which is localizable, and a wave, which is not localizable. When the quantities describing the state of a system are replaced by operators, the commutation relations of these operators have as their implications so-called *uncertainty relations* (formulated by Heisenberg), which show that certain pairs of quantities cannot have exact values simultaneously. For example, if one tries to determine exactly a local coordinate, the corresponding component of momentum remains necessarily uncertain, and vice versa. Such pairs of quantities are mutually complementary. The use of complementary pictures and concepts is necessary in the description of the atomic world, and the formalism of quantum mechanics reflects this fact (e.g., *Bohr* 1985, p. 108).

The idea of 'mutually exclusive' pictures has recently become problematic because it has been shown, for example, that *both* wave *and* particle properties can be necessary in understanding the same experiment. (The double-prism experiment, cf. *Ghose* 1992a & *Ghose* 1992b. See also *Löfgren* 1994.) This sharpens still more the idea of complementarity because it shows that an object in the same experiment can have both particle and wave properties; therefore one cannot say that these contradictory properties would appear only in different 'mutually exclusive' experiments. This shows that the objects of the microworld are contradictory to themselves which is an expression of the irrationality of reality (*Laurikainen* 1992). From the point of view of rational description one can say that *the concept of an 'object' becomes obscure on the microlevel.*

This may be the main reason for the criticism against the concept of complementarity: If one presupposes that all concepts must be defined with an accuracy required in exact sciences then the concept of complementarity cannot be defined because it implicitly presupposes irrationality. The complementarity of reality is a concept belonging to the realm of metaphysics. It and the irrationality of reality contain, in fact, the idea of the *transcendence of reality*. Bohr, who disliked metaphysics, avoided speaking of reality and realism for exactly such reasons.

The essential difference between the philosophies of Bohr and Pauli concerns the role of the observer in quantum theory. Especially in the 1950s, Bohr began to emphasize the objectivity of science to such an extent that he claimed the observer to be, even in atomic theory, a 'detached observer', i.e., the observer can, in principle, be neglected when describing an observation; the observer can be represented as a measuring instrument which undergoes a purely physical interaction with the object. In this way the introduction of psychic viewpoints to the analysis of observations is avoided.

Bohr thought that psychic viewpoints would bring in subjectivity which is not allowed in an objective science. In a carefully formulated article from the 1950s Bohr writes:

Entscheidend ist, dass in keinem Fall die geeignete Ausweitung unseres begrifflichen Rahmens eine Berufung auf das beobachtende Subjekt in sich schliesst, was eine eindeutige Mitteilung von Erfahrungen verhindern würde. (*Bohr* 1985, "Atomphysik und Philosophie. Kausalität und Komplementarität", p. 110.)

This view of Bohr's (together with the criticism to which Einstein subjected the Copenhagen interpretation) has caused psychic considerations to be continually excluded from the analysis of observations in atomic physics.

Pauli remarked to Bohr that the idea of a 'detached observer' is not compatible with Bohr's original description of complementarity; Bohr has often emphasized that in the great drama of existence we are not only spectators but also actors (*Bohr* 1985, e.g., pp. 18–19, 62). According to Pauli, the psyche of the observer is inseparably mixed with the description of the observed object because he plans the experimental arrangement and interprets the results of the experiment, and this of course influences the way the results have to be described. Bohr has said that quantum mechanics does not describe the atomic world itself (i.e., reality) but only helps us to describe the observable 'atomic events'. In the 1950s the 'Copenhagen people', especially Heisenberg as well as Born in Göttingen, used the phrase that quantum mechanics describes *our knowledge* of the atomic world, it does not reach the atomic world itself. (*Heisenberg* 1955, p. 12 & p. 18. *Born* 1983, p. 298.)

This means that the consciousness of the observer must be included in the analysis (*Pauli* 1994, e.g., pp. 40–42, 151 ff.). Pauli in fact considered observation as an 'interaction' between the consciousness of the observer and the observed object. However, this 'interaction' should not be understood as a causal interaction; it is a mutual influence between the 'world of spirit' and the 'world of matter' of Descartes, which so far has not at all been taken into consideration in natural science. Here we run into the *psychophysical problem* which Pauli found to be the most important problem of our time (*Pauli* 1994, pp. 153–159. *Laurikainen* 1988, pp. 57, 101 ff. and 141–142).

Unus Mundus

According to Pauli, the situation which we meet in the analysis of observations in quantum mechanics forces us to *abandon Cartesian dualism*. Pauli called the idea of *psychophysical parallelism* created on the basis of this dualism a *spiritual cloud of fog* in Western thought (*Laurikainen* 1988, p. 141) because the idea of 'parallelism' remains without scientific motivation and because during the last few centuries it has resulted in a sharp distinction being made between natural sciences and humanistic sciences as well as between science and religion. Heisenberg has also (e.g., in his book *Physik und Philosophie*, 1959) emphasized that atomic physics demands a new way of thought in place of Cartesian dualism (*Heisenberg* 1959, p. 66).

For Pauli, the complementarity of quantum mechanics supplied a model for approaching the psychophysical problem. He thought that physical and psychic phenomena are mutually complementary in the same way as the wave picture and the particle picture in atomic theory. Discussions and correspondence with C.G. Jung concerning depth psychology gave additional weight to this deliberation (*Pauli* 1994, article 17). Their common view, which was a result of a quarter of a century of discussions between them, was the idea of an abstract world, *Unus Mundus*, which expresses itself for us as physical phenomena and psychic experiences (*Arzt et al.* 1992). (The term unus mundus was borrowed from the German physician and alchemist Gerhard Dorn who lived in the 16th century.) To this concept Pauli added the idea of the complementarity of its physical and psychic aspects.

Pauli's complementary concept of reality opens a new possibility for approaching the psychophysical problem. It is a view which abandons both materialism and idealism and which presupposes that the 'matter' and 'spirit' of Descartes exist only as a whole which is simultaneously both 'matter' and 'spirit'. In fact, it is best to deviate from Cartesian terminology and call the 'material' aspects of reality *physical* and 'spiritual' aspects *psychic*, whence the abstract reality itself is psychophysical. (If we call this psychophysical reality spiritual – which thus is experienced as psychic and physical phenomena – this corresponds to some old conceptions in which there are three elements in a human being: body, soul, and spirit, the essence of man being spirit.)

In 1946 Jung also expressed the view that reality has a *psychoidic*, i.e., simultaneously psychical and physical, nature (*Pauli* 1994, p. 156).

In order to understand abstract, psychophysical reality, one has to notice the unconscious functioning of the psyche. Jung's depth psychology contains the important conception of the *collective unconscious* (*Pauli* 1994, pp. 157–159), which comprises everything that can potentially become conscious. The collective unconscious especially contains certain instinctive abilities which guide the functioning of our psyche, both the unconscious shaping (gestalting) processes of the stimuli arriving from the outer world and our conscious thinking processes. When describing these processes Jung introduced the old concept of *archetype* (which played an important role, e.g., in Neo-Platonism); it underwent considerable changes during the years and discussions with Pauli probably influenced this development (*Pauli* 1994, pp. 156–159). The concept of archetype in any case fitted very well with Pauli's views concerning the psychophysical world.

The experience of the tremendous influence which mathematics has on the development of physics had a strong effect on Pauli's considerations. Mathematics is a creation of the human spirit, a characteristic expression of our psyche. The fact that mathematics is applicable in such a wonderful way to the description of the order in the material world (especially of natural laws) is, according to Pauli, an expression of a *cosmic order* which is psychophysical, beyond the separation of the physical and the psychic (*Laurikainen* 1988, e.g., p. 21. *Pauli* 1994, e.g., p. 34). Therefore the idea of a psychophysical reality fit well into Pauli's worldview, and for him Jung's archetypes were structural elements in this psychophysical world. On this basis he hoped to find a way of describing the structure of psychophysical reality. That reality is, it is true, unattainable to rational description (i.e., it is transcendent), and its description is only possible in the sense of metaphysics: by using similes and analogies. Pauli wrote, in fact, that for him reality is *symbolic* in its very nature (in the same sense as Jung used the word 'symbol'). Only observed events are concrete (*Laurikainen* 1988, p. 21. *Pauli* 1994, articles 1 and 17; more in *Pauli & Jung* 1992). An observation does not belong to the symbolic reality; it is an 'actualization' of one 'potentiality' belonging to reality.

It is impossible to define the concept of archetype. Their expressions are, according to Jung, archetypal images and symbols which are used in religions and mythologies and which have a strong emotional energy charge. Jung has shown that these symbolic figures have very similar symbolical meanings across different cultures. According to him, they are, therefore, expressions of the structure of the objective 'collective unconscious'. Archetypes themselves are hidden in the depths of the unconscious, and one can only guess their meaning on the basis of the conscious figures and ideas which they produce.

This kind of view of archetypes matched well with Pauli's idea of the cosmic order which is beyond our experiences concerning the physical and the psychic. Instead of symbolical figures, Pauli saw expressions of archetypes in mathematical concepts: in natural integers which he thought of also in the sense of the number mysticism, in the concepts of the mathematical theory of groups, and in symmetries.

The psychophysical reality of Pauli and Jung has in the recent years become an object of increasing interest. Since their correspondence has now been published (*Pauli & Jung* 1992), one can expect that interest in Pauli's complementary conception of reality will continue to increase. An analogy to the idea of complementarity can be found in Western philosophy in the idea of dialectics (for instance in Hegel's philosophy and in dialectical materialism). However, the discreteness of quantum theory (symbolized by Planck's constant h) gives to this "idea of opposites which together form a whole" (*Pauli* 1994, e.g., article 1), the depth which really opens new perspectives for ontology. Complementarity is created by a combination of discreteness and continuity. The whole formed by these opposites results in reality which contains both rational and irrational aspects. This leads to a new view concerning the relation between the physical and the psychic. It is to be hoped that the further development of the ideas of Pauli and Jung will have a fruitful influence on the direction of the Western thought, and will finally bury materialism which has in a dangerous way directed Western thought to narrow and untenable utopias.

7. On the Meaning of Complementarity

In his 'last interview' (with Thomas Kuhn, 17.11.1962), Bohr criticized philosophers because they had not made it clear for themselves what complementarity in fact means:

I felt ... that philosophers were very odd people who really were lost because they have not that instinct that it is important to learn something and that we must be prepared to learn something of very great importance. ... There are all kinds of people, but I think it would be reasonable to say that no man who is called a philosopher really understands what one means by the complementary description. ... They did not see that it was an objective description, and that it was the only possible objective description. (Quoted according to *MacKinnon* 1982. The interview is kept in the Niels Bohr Archive, Copenhagen. *Sources for the History of Quantum Physics*: Bohr, Interview 5.)

I will try to elucidate, with the aid of some remarks, why I think that this criticism by Bohr is well-founded. I also try to show that Bohr's attempt "to learn something of very great importance" requires that the 'epistemological lesson' given by nature has to be understood still more profoundly than Bohr himself was willing to do: One should not exclude the discussion of the ontological implications of quantum mechanics.

1. There are two basic conceptions in the Copenhagen interpretation which are seldom truly understood. a) If the particle picture and the wave picture are associated with one and the same object, the theory becomes internally controversial and therefore useless. b) The need for complementary ways of description has to be seen as an expression of the very nature of reality, so that there is no reason to expect that this feature would be eliminated by a future development of the theory.

Because of the fact mentioned in a), Bohr pointed out that the wave picture is not associated with a particle but these pictures are related to given experimental situations and, therefore, the description of a microsystem has to be understood as only symbolical:

Es muss jedoch erkannt werden, dass wir es hier mit einem rein symbolischen Verfahren zu tun haben, dessen eindeutige physikalische Deutung letzten Endes den Hinweis auf eine vollständige Versuchsanordnung fordert. (*Bohr* 1985, p. 109. Cf. *Bohr* 1963, p. 5.)

If complementarity is understood in this way, the idea – usually associated with realism – of the world as composed of individual objects is obscured. Bohr said, in fact, that he did not understand what 'reality' can mean on the microlevel. The excellent results of quantum mechanics concern atomic *phenomena*. In addition to the quantum mechanical description of phenomena, it is only possible to state that the microworld differs from the macroworld in that one needs there complementary descriptions, which 'exclude each other'. Realism, in the traditional sense of this word, becomes problematic.

The view described in b) concerns the 'completeness' of quantum mechanics, and this is the issue debated by Bohr and Einstein in their controversy. Einstein was convinced that there are features in physical reality which quantum mechanics does not describe at all. This is, according to his view, the origin of the 'paradoxical features' of quantum mechanics, among other things, of statistical laws, which must be eliminated when the theory develops. In the Copenhagen interpretation, it is instead presupposed that complementarity and statistical laws will never be eliminated from the theory because they reflect the very nature of the phenomena.

It is important to notice that determinism is, in classical physics, expressed in the form of differential equations, and it is understood that the most complete laws are found 'im Kleinen'; by investigating smaller and smaller details. The fact that determinism is found impossible just in microphysics is an indication of the incorrectness of the idea of determinism itself.

2. There has been some discussion recently about the characterization of complementary ways of description as 'mutually exclusive'. For example, in the so-called two-prism experiment both particle and wave properties are needed in the interpretation of the same experiment. (*Ghose et al.* 1992a, pp. 403–406 and *Ghose et al.* 1992b, pp. 95–99.) In fact, this is the case even in a diffraction experiment of radiation if one is able to follow the formation of a diffraction pattern by using a very low intensity in the incoming beam, so that the prints made by individual particles on the film can be clearly separated. (See figure on the frontispiece of Part I.) In both cases one can observe signs of localizable 'particles' in the detector (in the diffraction experiment prints on the film), but their *distribution as a whole* can only be explained with the aid of the wave picture.

Bohr's claim that the particle picture and the wave picture are never needed simultaneously, is, thus, not quite correct. The new experimental techniques show that this statement used by Bohr for describing the complementarity idea is a little inaccurate. However, even more clear now is that the concept of an 'object' becomes obscure on the microlevel. This has importance with respect to ontology.

3. *Thus, waves always presuppose sufficient statistics.* Here one is forced into the statistical interpretation. The idea of an object which 'simultaneously is both wave and particle', however, is very vague, and therefore Bohr said that quantum mechanics describes the microworld only symbolically.

Pauli emphasized the same thing by starting from the fact that laws in microphysics are in principle statistical. In the individual events one then meets the irrationality of reality: There are features in reality which cannot be completely described by any rational theory. Reality is then found to be transcendent, not completely describable by rational theories. One has to abandon traditional realism, which is associated with the correspondence theory of truth, because it is not possible to define the reality which should be the basis for the comparison of theories.

This result, that one must abandon realism in the traditional sense of the word, is an important implication of the Copenhagen interpretation; experimental results which created quantum mechanics, necessarily lead to this conclusion. (A thorough discussion of the situation which leads to statistical laws and the crisis of realism is given in *Pauli* 1994, articles 2 and 3.) Therefore it is strange that, since the 1960s (after the death of Bohr), the new attempts to interpret quantum mechanics (as well as the quantum theory of measurements) have all been based on realism. Even those who accept the Copenhagen interpretation base their thinking on realism, although this, according to what has been said above, is a logical mistake. This is a consequence of the pragmatic attitude among physicists.

4. Because of this situation, Born and Heisenberg have stated that quantum theory does not describe reality but only *our knowledge of reality*. (See quotations on pp. 50, 51.) They came to the conclusion that Cartesian dualism cannot be maintained in microphysics; in the description of the microworld, the concept of *consciousness* is necessary. Pauli has described the same view more in detail in his epistemological essays (*Pauli* 1994).

This is perhaps the reason for the attempts to describe microphysics according to realism, although the analysis of corresponding experiments shows that this is impossible. Bohr has certainly strengthened this tendency by warning that the objectivity of physics is in danger if psychic matters are mixed with physics; in his last years Bohr very strongly emphasized the requirement of objectivity in science. (E.g., *Bohr* 1985, article 8. This article was written after Fock's visit mentioned in Section 9.) Bohr has been inconsistent in this respect because the facts, from which the idea of complementarity has arisen, indisputably show that a conscious observer is needed in the interpretation and his choice of the method of observation has an essential influence on the experimental 'facts' obtained. I would like to say that Bohr here wished to adapt himself to the criticism of philosophers which he, however, in the statement quoted in the beginning criticized. It is as if he stopped halfway when emphasizing that "we must be prepared to learn something of very great importance". Bohr did not venture to step over the border inherent in Cartesian dualism because then it would be necessary to introduce psychic matters into physics. The strong political power of the materialistic ideology perhaps had here a certain influence on Bohr's judgement. (See Sect. 9 below.)

The clearest conclusions from the situation were drawn by Pauli, who together with C.G. Jung tried to understand the psychic matters which are basic for shaping (gestalting) the picture of the world. Then he was forced to think also of the unconscious functioning of the psyche because this is the basis for our conceptions. (*Pauli* 1994, article 17. The unconscious is a central theme in *Pauli & Jung* 1992.)

5. If matter waves – i.e., the state function of quantum mechanics – are interpreted as describing our knowledge of reality, the paradoxes of quantum mechanics disappear. One runs into paradoxes only if realistic views are associated with quantum mechanics. The heart of all paradoxes is the reduction of the state function due to an observation. However, if the state function describes our knowledge of the situation, it is natural that it is changed in an unpredictable (irrational!) way by an observation, since our knowledge of the situation changes.

A state function interpreted in this way is not a wave propagating in the outer world, although it is formally described as a wave. It is a concentration of the knowledge we have of the system, describing its statistical mean behavior – when statistics are sufficient – in the situation defined by the experimental arrangements. Only this mean behavior is describable by laws. In individual events there appears dispersion (scattering) which cannot be governed theoretically. The mean behavior described by the state function is in microphysics the only regularity describable by laws, while we have to abandon the deterministic description of individual events. This is the idea of statistical causality. (Pauli describes statistical causality in *Pauli* 1994, article 3; however, he proposed the name 'statistical correspondence' instead of the generally used 'statistical causality'.)

In the limiting case of very massive systems the statistical laws of microphysics are reduced to deterministic laws of classical physics (Bohr's correspondence principle). Pauli interpreted the lesson given by microphysics in a more general sense than just concerning the microworld. He considered the very idea of the deterministic description of the world to be unsatisfactory. We meet, on the macrophysical level, many other kinds of phenomena than merely physical ones, and after the experiences in microphysics concerning the nature of causality we have reason to think that these are not governed by determinism but rather in the way characteristic of statistical causality. This implies that we must acknowledge irrational aspects in reality, in addition to the rational ones described by the scientific theories.

This irrationality contains the *creative element* of reality, because it contains 'choices' between different possibilities allowed by the rational statistical laws, and these 'choices' cannot be predicted on the basis of what existed before the 'choice'. 'Choices' create something new within the limits stipulated by the conservation laws.

6. *A matter wave can be characterized as a symbolic shape (gestalt) in the consciousness*, which makes it possible to govern, in a certain sense, a whole

consisting of a great amount of similar phenomena. Of course, we believe that it reflects certain features in the 'reality-itself', but this is a metaphysical claim. The fact that the predictions derived from a 'wave' according to the probability interpretation are so reliable, as they have been shown to be, justifies the claim that the wave 'is real'. But critically speaking, the wave describes only our knowledge of reality in the sense that it makes possible reliable statistical predictions concerning a great number of 'similar' events. (What 'similar' means here, is defined by experimental arrangements.)

The state function is holistic by nature. The state function of a system of particles describes this system *as a whole;* it does not make a distinction between individual similar particles. It is a shape (gestalt) created by the psyche in order to describe a group of 'particles' as a whole, which in general has properties quite different to these individual 'particles'. The wave function makes quantum mechanics a holistic theory.

In an experiment involving 'similar' and 'similarly prepared' particles the state function combines individual *events* into a whole – a holistic shape. Often this whole can be seen very concretely as a regular structure on a film, for example. It is possible to speak of a 'wave associated with an individual particle' only in the sense that these 'particles' have the property of behaving *statistically* in the way described by the wave function. The 'wave' can, in fact, only be demonstrated by using a very great number of 'similar particle events'.

Such 'paradoxes' as the EPR results thus acquire a natural explanation. (As a description of the EPR problems and the Bohr–Einstein controversy see, e.g., *Shimony* 1992a.) 'Matter waves' just give a means for governing certain experimental situations in the statistical sense. A matter wave does not 'propagate' in the outer world; it is only a shape in the human consciousness. The 'non-locality' and the 'instantaneous predictions' found in these experiments are understandable because the functioning of the psyche is not restricted by the local and temporal conditions characteristic of the physical world; the EPR correlations follow from a shaping process which takes place *in the consciousness.*

Space and time are shaping forms when searching for invariant structures in the outer world. Causality is a shaping form for finding invariances in 'similar events'.

7. The materialistic conception of reality has distorted the interpretation of the wave function in a very strange way. Since it was found (finally in the 1950s) that the Copenhagen interpretation necessarily leads to the introduction of consciousness into the ontology, there appeared a strong tendency to seek a more realistic interpretation acceptable from the point of a materialistic conception of reality among the young generation educated during the war or after it. This tendency has, in fact, become quite dominating among scholars who actively participate in the discussion of the philosophical questions. There are plenty of them who find the Copenhagen interpretation unsatisfactory. They are waiting for the development to lead to a change in the interpretation of quantum theory, or

perhaps in the theory itself, making realism indisputably possible and simultaneously eliminating the 'paradoxes' burdening quantum mechanics in its present form. This even seems to be the ruling attitude - with the exception of the great majority which is not interested in the philosophical questions but formally accepts the Copenhagen interpretation. There is no scientific motivation for this realistic/materialistic belief of the younger generation, as far as I understand. It is based on the belief that the development will take place in a direction which in fact seems more and more improbable as all the new approaches can be found unsatisfactory. One should, at least, show how the reduction of the state function and the other paradoxes encountered in the realistic interpretation can be overcome.

The introduction of consciousness has become a curious ghost for people accustomed to materialistic paths of thought; its nature is misunderstood in a very strange way. In his carefully written books concerning the foundations of quantum mechanics, Franco Selleri, for example, interprets the state function as a real element of the outer world, which can only change due to a physical cause (*Selleri* 1990, pp. 162-167). According to him, the consciousness can play a role in an observation only if one can show that it can have observable effects on physical phenomena (i.e., some kind of psychokinetic influence). The consciousness is understood to have causal influences on the (real) state function.

For Pauli and the other physicists of the old generation, who found the introduction of the consciousness to be necessary in the interpretation of quantum mechanics, its role meant realizing a change in the outer world and drawing consequences from the observed facts – a shaping process. It is not possible to measure such psychic processes by using physical measuring methods. The consciousness (or more generally, the psyche) is not in any causal interaction with the object of an observation. There must be some kind of 'interaction' but this must be something like the 'pre-established harmony' of Leibniz – i.e., the mutual order of monads, while causality belongs to the internal phenomena of a monad. Correspondingly, Schopenhauer spoke of the 'underground connections' between phenomena, and C.G. Jung proposed a new ordering principle of phenomena called *synchronicity,* alongside causality (C.G. Jung: "Synchronizität als ein Prinzip akausaler Zusammenhänge" in *Jung & Pauli* 1952).

The 'interaction' between the consciousness and the observed system means an 'interaction' over the 'insurmountable borderline' between matter and spirit. There must, of course, be some 'interaction' between the world of spirit and the world of matter in order for experimental research to be possible. Pauli proposed that this 'interaction', i.e., the psychophysical problem should be approached on the basis of the idea of complementarity, using quantum mechanics as a model: a complementary theory of 'particles' and 'waves'. On the basis of the remarks made above one could imagine that 'particles' could be compared with matter and 'waves' with the psyche, especially as 'matter waves' have been said to describe 'knowledge'. Just as the microworld has been found to be so abstract that its realistic description, in the traditional sense of the word, is not possible, so is the reality formed by 'matter' and 'psyche' unquestionably transcendent in the sense that we cannot reach it by rational theories.

In quantum mechanics, complementary quantities are characterized by a certain non-commutative algebra, i.e., certain *commutation relations*, which are the formal heart of quantum theory. How could we find something analogous in the psychophysical reality?!

8. Personally I think that *it is not possible and there is no need to search for a new ordering principle alongside causality*. One has to abandon still more consistently the patterns of thought inherited from determinism. Schopenhauer speaks, in fact, of an *irrational will* when discussing the 'underground connections' between phenomena. One should understand the 'freedom' inherent in statistical causality to be real freedom, not bound by any additional ordering principles, apart from the statistical law which represents causality. This 'freedom' is the creative element of reality: It means truly free 'choices' – remaining, however, within the framework of the statistical law.

Irrationality seems to hide an explanation for the obvious purposefulness which is particularly characteristic of living organisms. Reality seems to include an aspect which reveals itself as 'free choices' in individual events. In inanimate nature this aspect can in general be described as 'chance', but in living organisms it is difficult to avoid speaking of purposefulness. Life can even be characterized as some kind of awareness of the circumstances, with purposeful choices corresponding to the situation in the environment.

It is interesting that these features appear in a reduced form in the state function of quantum mechanics. 'Circumstances' are defined by the experimental arrangements which appear as boundary conditions: The state function is a solution of the Schrödinger equation which fulfils the boundary conditions corresponding to the experimental situation. On the other hand, the law which in this situation 'regulates the choices' is formally contained in the Schrödinger equation (more definitely, in the Hamiltonian operator describing the system in question). Individual choices are free, but the statistical law shows that there is a 'will behind the choices' which 'takes care of' the right holistic behavior corresponding to the statistical law; this behavior is characteristic of the systems in question (described by the Hamiltonian mentioned above).

This 'will behind the choices' thus has both rational and irrational aspects.

9. At first Bohr often stated that "causality is not valid in microphysics." When the Soviet Academician V.A. Fock in 1957 visited Copenhagen, he remarked to Bohr that by making small changes in the wording it is possible to avoid the 'unnecessary' controversy between the Copenhagen interpretation and the materialistic philosophy. He pointed out that there are laws even in microphysics although of a different kind to those of macrophysics. He proposed that one could in microphysics speak of *probabilistic causality*. (See *Laurikainen* 1985b and *Laurikainen* 1988, pp. 57–62.) Afterwards Bohr, indeed, avoided the phrase that

causality is not valid in microphysics; he emphasized, instead, that causality must be generalized to the 'framework of complementarity'. (See, e.g., *Bohr* 1985, article 8.) Practically this generalization means the idea of statistical causality in the sense we have discussed it here.

From the very beginning, Bohr understood complementarity as a generalized causality. Pauli has described this more clearly: Instead of complementarity he emphasizes the statistical (probabilistic) laws as the most characteristic feature of quantum mechanics. He has also analyzed the situation more profoundly from the philosophical point of view. Bohr obviously had a strong antipathy to all metaphysics, and this prevented him from discussing the ontological problems created by quantum mechanics and the role of psyche in shaping the picture of the world. Pauli has very clearly described the nature of statistical causality, including its ontological implications; according to his view reality has both rational and irrational aspects.

Here I have mostly described Pauli's views, perhaps excluding the remarks at the end of Section 7 and in Section 8. At issue is, in the first place, a critical analysis of the philosophy implied by quantum mechanics, including ontological implications. This can be most successfully made on the basis of Pauli's views.

10. The reason for the difficulties that philosophers encounter in acknowledging the idea of complementarity is probably that they are not accustomed to adapt their thinking to empirical facts. Nature has, in atomic research, given us a lesson which cannot be assigned to any existing philosophical 'compartments'. This is a situation which seems to be hard to accept: that "we must be prepared to learn something of very great importance."

An important aspect of this lesson is that *reality remains trancendent*, in principle, but in spite of this it is possible to know something of reality - something that is so certain that it is possible to base practical actions, even technology, on this knowledge. One has begun to speak of empirical philosophy: Some problems concerning knowledge and the nature of reality can, in fact, be illuminated most clearly with the aid of experiments.

For the theologians the new orientation can, perhaps, be still more difficult. Especially in the Protestant theology the special revelation given in the Bible has been raised to a unique basis of belief and the general revelation which is inherent in the Creation has been almost forgotten. This attitude is, perhaps, most categorical in Calvinism. This is an inheritance from the discussions, in the late medieval age, concerning God's influence on phenomena in the created world – discussions which also had an influence on the birth of empirical science. (*Työrinoja* 1994. *Burtt* 1980.) These discussions should now be continued on the basis of 'empirical philosophy'.

In microphysics we have learned that the universe is not a machine but the arena of a continuous creation. The invariant reality of Plato was, in the 17th century, replaced by a world of changes. However, the changes were found to be governed by invariant natural laws. Now even the idea of natural law needs an essential change. The idea of determinism is found untenable.

At issue is the concept of causality. Physics has been forced to replace deterministic laws by statistical (probabilistic) laws. This is a revolution rocking the foundation of our picture of the world – although few physicists have realized the philosophical importance of this change. Realism in its traditional form is still the ruling attitude in physics. It is presupposed that we can describe in physics an independent reality – independent of any human observations – and that one can even use mathematics in this description. Practically this means that one has not yet abandoned the deterministic description of the world.

The idea of statistical causality means a change which should influence even the attitude of the theologians. The question of predestination, especially, is inseparably bound up with the idea of causality. One should learn to think that causality only concerns the mean behavior. In individual events there is always an aspect of freedom which people have so far ignored, both in science and in theology.

When thinking of human behavior this is a natural idea. Man has free will, but he must adapt himself to the general laws which determine the mean development – independently of our individual decisions. In his free choices man has responsibility for their consequences.

If we accept the possibility of the general revelation – which belongs to the Lutheran theology, in principle – this should mean that we can learn something of God also by studying the material world. The Creation is like an image of its Creator. In this sense the idea of complementarity emphasizes, in the first place, that *God is transcendent and simultaneously rational and irrational.* Everything we say about God has to be understood symbolically – we can only use similies and analogies. God's rationality is reflected in causality which is the basis for our comprehension of nature. However, causality does not mean determinism. Laws inherently contain freedom which seems to be completely independent of any laws. This is the lesson of microphysics.

This can perhaps be illuminated by comparing the deterministic idea of classical physics with the Old Covenant, where laws and the justice of God are primarily emphasized. In the New Covenant, however, Grace is more important than Law. It means freedom independent of any laws. This reflects the irrationality of God. God is revealed both in the rationality of the universe, in its invariant basic features, and in the irrational freedom which makes continuous creation possible. This is also the basis for the free will of man. God is the origin of both Being and Becoming.

Science has one-sidedly emphasized the rationality of the world, but this leads to a distorted culture where expressions of spirit and the possibility of creation are eliminated. This results in a picture of the world as a deterministic machine. Unfortunately this tendency still dominates in science.

This chapter was published in Finnish in *Teologinen aikakauskirja* 1/1996 (Helsinki).

8. On the Criticism by Natural Scientists

This chapter is a protest against the way philosophical questions are dealt with in natural sciences. The present expert system makes the criticism of scientism and materialism almost impossible. Scientists who specialize in philosophy would be needed in faculties of natural sciences. Some examples are given of the supercilious style in which one tries to silence remarks concerning the spiritual aspects of reality.

Physics and Philosophy Today

At an open discussion at the University of Helsinki on December 8, 1994, Professor Stig Stenholm, in his introductory talk, quoted a passage from Steven Weinberg´s book *Dreams of a final theory* (*Weinberg* 1993, p. 134):

I know how philosophers feel about attempts by scientists at amateur philosophy. But I do not aim here to play the role of a philosopher, but rather that of a specimen, an unregenerate working scientist who finds no help in professional philosophy. I am not alone in this; I know of no one who has participated actively in the advance of physics in the postwar period whose research has been significantly helped by the work of philosophers. I raised in the previous chapter the problem of what Wigner calls the 'unreasonable effectiveness' of mathematics; here I want to take up another equally puzzling phenomenon, the unreasonable ineffectiveness of philosophy.

This quotation gives a good example of the strange situation which is characteristic of the relations between physics and philosophy at the end of the 20th century. Physicists expect, like Weinberg, that philosophy should produce new physical theories in order to 'be effective'. Weinberg does not seem to think at all of the possibility that philosophy could change the direction of the endeavor in physics – which cannot happen in a year or two. And so he constructs, on the basis of a (perhaps unconscious) materialistic belief, cosmological theories without considering that Pauli, for instance, has made philosophical remarks which shake the basis of this belief. The idea seems to be quite strange to physicists today that they should also have the responsibility for the philosophical foundations of their most general theories, so that one does not make the mistake of applying physical methods to questions where their use cannot be properly justified. Scientific methods are not a sufficient guarantee for the reliability of the results.

I know that here I am criticizing a Nobel laureate. I dare to do so because another Nobelist has emphasized that "they went a little too far in the 17th century." If one really understands the nature of the decline of determinism and the irrationality of reality which it implies, Weinberg's description of the first three minutes of creation *(Weinberg* 1978) begin to look like a misuse of scientific methods.

John C. Eccles – a Nobel laureate, too, by the way – in his book *How the Self Controls Its Brain (Eccles* 1994, preface p. IX) has described the general attitude of natural scientists today as follows:

In the philosophy of the earlier decades of this century we were immersed in the long dark gloom of behaviorism, Ryleanism, logical positivism, Skinnerism, and so on. I agree with the appraisal of Roger Sperry as expressed in the extracts in Chapter 3.10 and 3.11. Since the 1950s Sperry and I have challenged this materialist interpretation of brain science, but we have been largely ignored. Materialists remain as dominant as ever because they are devotees of a dogmatic belief system which holds them with a religious-like orthodoxy, as expressed by Edelman's materialist metaphysics in Chapter 3.5.

I think that this orthodoxy has now been most profoundly shaken by quantum physics which forces us to re-think the basic beliefs concerning reality and the foundations of human knowledge. Unfortunately, physicists are devoted supporters of the materialistic orthodoxy. They ignore philosophical problems because they are in a hurry to get results. Therefore, the ideas concerning the foundations of quantum physics of Wolfgang Pauli lie forgotten in the archives. People find them 'mystical'. They could occasion criticism with respect to the results which help physicists to get support from society!

The problem is that physics is recovering the *soul* which during the last three centuries has been more and more forgotten. Some very critical physicists – who do not think lightly of philosophy – have begun to think that we should not ignore the possibility that reality contains something that simply cannot be described by the methods of the present science, something that cannot be measured.

Look how G.M. Edelman, whom Eccles mentioned above, confines the dominion of scientific research *(Eccles* 1994, p. 33):

Any adequate global theory of brain function must include a scientific model of consciousness, but to be scientifically acceptable it also must avoid the Cartesian dilemma. In other words, it must be uncompromisingly physical.

And further *(Eccles* 1994, p. 34):

Scientific epistemology must confront the issue of consciousness in terms of evolution, development, brain structure, and the physical order as we know it. If the confrontation is to remain in the scientific domain, a dualistic solution or any form of Cartesian empiricism cannot be countenanced, [being] often accompanied by what might be called Cartesian shame.

You see, Descartes accepted the idea of the soul dwelling in the 'world of spirit' (res cogitans) which existed alongside the 'world of matter' (res extensa).

Because of Edelman's conception of science Eccles writes (*Eccles* 1994, p. 34):

This prevailing critical atmosphere bleakly recalls an inverse Inquisitorial degradation!

While writing this, I received an issue of *Time* (July 31, 1995) which was devoted to research into consciousness. In it, the psychophysical problem is described as one of the burning issues of today, understood in the spirit of Edelman's restrictions. At the end of the coverage there is, however, a cautious mentioning of the possibility that eventually science can become forced to confess that there is something that can be called the soul.

The Role of Consciousness in Atomic Theory

It has been pointed out above that the materialistic conception of reality makes the role of consciousness in quantum mechanics very difficult to understand. Let us think about this question in a little more detail.

In Italy, research on the foundations of physics in the spirit of materialism seems to have a very strong position. Contacts with Soviet physicists may have had an influence on this. As an example of the influence of materialism on the interpretation of quantum mechanics, we can take Franco Selleri's book *Die Debatte um die Quantentheorie* (*Selleri* 1990). First I wish to state, however, that I have found Selleri's works in general very carefully written, and it is precisely for this reason that his remarks about consciousness are a good example of the influence of the philosophical basic view on the work of a scientist.

The last part of the book has the title "Experimentelle Philosophie," and it includes a chapter entitled "Eine Rolle für das Bewusstsein?" (A role for the consciousness?). In this chapter, Selleri presents grounds for his opinion that the state function cannot represent only the knowledge of the observer: this would require an influence of the consciousness on the physical phenomena in a way which has not been observed.

Selleri presupposes that it is possible to speak of the state of a microsystem independently of any observations. Thus he ignores the implications of Bohr's 'quantum postulate' (the introduction of Planck's constant into atomic theory) concerning microsystems and microphenomena. Selleri presupposes that all details of phenomena, even the reduction of the state function, must have causes which are analyzable by scientific methods. The concept of consciousness can only be introduced if it can be in causal interaction with physical systems: It must have measurable influences.

A passage from Selleri's book may illuminate this view (*Selleri* 1990, p. 165):

The above-described discussion concerning a measuring process follows partly Wigner's description: "The modified wave function [after the reduction] is, furthermore, in general unpredictable before the impression gained at the interaction has entered our conscious-ness: it is the entering of an impression into our consciousness which alters the wave function because it modifies our appraisal of the probabilities for different impressions which we expect to receive in the future." (Cf. *Wigner* 1967, p. 175.) From these premises Wigner draws the conclusion that "it will remain remarkable, in whatever way our future concepts may develop, that the very study of the external world led to the conclusion that the content of the consciousness is an ultimate reality." (*Wigner* 1967, p. 172.) Wigner concludes further that "we are called upon to construct a 'psychoelectric cell', in order to compose here a catchword."

N.B. The last sentence cannot be found in the article in question, "Remarks on the Mind-Body Question", *Wigner* 1967!

Because of his materialistic conception of reality, Selleri has completely misunderstood Wigner's meaning. Wigner wishes to point out that the reduction of the state function takes place in our consciousness because of a measuring result: because of this result we can form a new, 'reduced' state function and make new statistical predictions based on it. The state function is changed only when an observer becomes conscious of the result, not at the moment when a detector registers an observable result which everyone can see later. Heisenberg emphasizes exactly the same point in the citation presented on p. 38 above (*Heisenberg* 1959). An English translation of this passage reads:

Therefore, the transition from the 'possible' to the 'actual' takes place during the act of observation. If we want to describe what happens in an atomic event, we have to realize that the word 'happens' can apply only to the observation, not to the state of affairs between two observations. It applies to the physical, not the psychical act of observation, and we may say that the transition from the 'possible' to the 'actual' takes place as soon as the interaction of the object with the measuring device, and thereby with the rest of the world, has come into play; it is not connected with the act of registration of the result by the mind of the observer. The discontinuous change in the probability function, however, takes place with the act of registration [by the mind of the observer], because it is the discontinuous change of our knowledge in the instant of registration that has its image in the discontinuous change of the probability function.

Instead of 'results', Wigner speaks above of 'impressions' because he wishes to emphasize the essential role of consciousness/spirit in observation. In the second passage which Selleri quotes above, Wigner clearly supports 'idealism', stating that the psychic aspects of reality are primary, as compared with the material aspects. Pauli does not see any reason for jumping from the materialistic monism into a spiritual monism (idealism). As a representative of Western empirical science who has found materialism insufficient, he considers the physical and the psychic as equal basic aspects of reality. However, instead of dualism he recommends a new conception of reality where these basic elements are *complementary to each other*, in an analgous sense as this concept is used in quantum mechanics.

It is surprising that Stapp, in his book *Mind, Matter, and Quantum Mechanics* (*Stapp* 1993), quotes the above passage of Heisenberg twice (pp. 125, 177) but truncates it by leaving off the whole last sentence. This makes the role of consciousness quite unclear. Because Stapp especially emphasizes Heisenberg's ontological views, this seems to be an indication of a severe misunderstanding of Heisenberg's views.

A similar mistake has been quite common in the remarks of my physicist colleagues concerning consciousness. Although I have repeatedly pointed out that the role of consciousness does not concern any parapsychological (psychokinetic) influences, the introduction of consciousness has been rejected because it would mean that the observer can, by force of his consciousness, influence the results. The patterns of thought dominated by materialism and naive realism seem to bind thinking so strongly that it is impossible to understand the role of consciousness in shaping (gestalting) our picture of the world.

Microphysics and Realism

In discussions concerning the limits of science which have taken place during the last few years, at the initiative of the Finnish Society for Natural Philosophy, it has become increasingly clear that the philosophy implied by the Copenhagen interpretation is no longer correctly understood. (See Report series of Research Institute for High Energy Physics, University of Helsinki, especially *Hämäläinen et al.* 1994.) It is mixed with a realistic conception of the microworld, and this leads to contradictions and paradoxes. These difficulties are mostly international because people in general combine quantum mechanics with the traditional realism. This section and the following one describe the situation, mainly on the basis of the material recorded in our discussions.

First let me repeat the philosophical lesson obtained from atomic physics according to the Copenhagen interpretation, as I understand it:

1) The non-zero value of Planck's constant implies a discontinuity of changes at the microlevel: phenomena are composed of 'quantum jumps', each of which is an indivisible whole leading from an initial state into a final state. A 'jump' or quantum phenomenon must be defined by macrophysical experimental arrangements which are needed for the investigation of this phenomenon. These experimental arrangements define the initial state of this quantum jump.

2) The final state of the jump cannot be predicted with the aid of any deterministic laws. However, the state function, which is defined on the basis of

the experimental arrangements characterizing the phenomenon in question, predicts the possible final states and gives statistical probabilities for each of them; the wave function also predicts the statistical dispersion of individual results in a series of events observed under the defining experimental arrangements.

This is the idea of statistical causality which replaces the deterministic causality of macrophysics.

3) The details of a quantum jump cannot be investigated, i.e., it is not possible to describe what 'happens' between the initial state and the final state. In order to investigate experimentally the details of a quantum jump, some new experimental arrangements are needed for the additional measurements, and this means that we are no longer speaking of the original phenomenon.

For this reason it is not possible to measure the value of the state function at different space-time points. The state function is a purely theoretical means for making predictions concerning the final state. The quantum-mechanical description of the microworld is, therefore, a purely symbolical (theoretical) description of possibilities (i.e., a probabilistic description). The reliability of this description can only be judged on the basis of the statistical predictions it gives concerning the final state.

This implies that the propensity of which Karl Popper speaks cannot be considered some kind of field, as he thinks (e.g., *Popper* 1992a, p. 93 ff. and *Popper* 1992b, p. 186 ff.): The strength of a field is in principle measurable at each space-time point, while propensity is not. Propensity is essentially equivalent to the wave function, and it does not in any way eliminate the problems which some people see in the statistical interpretation of the theory or in complementarity.

A. The Criticism by Raimo Lehti

Professor Raimo Lehti, a respected mathematician and astronomer, has written an extensive essay on the basis of the seminar held on May 16, 1994 and the additional material related to it (*Lehti* 1994). The theme of the seminar was the border between science and non-science. Lehti also participated in the seminar "Observation in Science" in Spring 1995. Thus, we have a fairly clear picture of his attitude, and because he obviously helds views that are quite widespread among natural scientists, his essay is worth discussing in some detail.

Lehti declared that he accepts naive realism. Thus it is clear that Lehti has not been interested in the problems of observation which are the center of interest here. His often sharp criticism of the philosophical remarks made on the basis of the Copenhagen interpretation is understandable if he has tried to comprehend them on the basis of naive realism: The result can only be a series of strange misunderstandings. Quantum mechanics was born when microphysical experiments were found to be in irreconcilable conflict with the prevailing idea of realism and it was necessary to try to find a completely new view of the nature of the physical reality. Lehti has decided to ignore such a possibility: Obstinately, he presupposes that traditional realism with the correspondence theory of truth must be absolutely accepted in science. No wonder he finds the epistemological and, even more, the ontological remarks made in this connection quite incomprehensible and impossible to accept. I think that Lehti represents here a common attitude among many natural scientists who do not express their views publicly.

Let us think, as an example, of the sharp remarks Lehti makes concerning the irrationality of reality (*Lehti* 1994, p. 102). Although I have again and again tried to describe this new feature of the conception of reality by discussing individual events in the case of genuinely statistical laws, Lehti still asks me to define what I mean by irrationality. It is an issue of a metaphysical change in the conception of reality which genuinely statistical laws imply. Lehti seems to require such an accuracy in the scientific language that all irrationality is absolutely excluded. "Desto schlimmer für die Tatsachen!" (Too bad for the facts!)

In addition, Lehti seems not to read what I write but interprets statements according to the terminology he is accustomed to using. My remarks then seem to be an inconceivable muddle. But this does not concern my texts only. In his essay (p. 102), Lehti quotes Pauli's description of reality in quantum mechanics. He gives a 'free translation' of Pauli's passages which have been presented in German in Ch. 5, p. 52 above (*Pauli* 1984, pp. 21–22). Lehti translates them abridged into Finnish, approximately as follows:

When concrete phenomena are (in observation?) fixed as actual, there appears in them an irrational aspect which has, as its opposite [counterpart], the verifications of the abstract system of possibilities with the aid of the mathematical concept of probability and the ψ-function ...

... The comprehension of the possibilities of natural events with the aid of quantum mechanics appeared to be a satisfactory framework to include in it also the actuality of the unique [the single event].

This translation sounds like nonsense. The passages are translated in the recently published English edition of Pauli's epistemological essays (*Pauli* 1994, pp. 46, 47) in the following form:

Observation thereby takes on the character of *irrational, unique actuality* with unpredictable outcome. Moreover, the impossibility of subdividing the experimental arrangement without essentially altering the phenomenon, brings a new feature of *wholeness* into physical happening. Contrasted with this irrational aspect of concrete phenomena which are determined in their *actuality*, there stands the rational aspect of an abstract ordering of the *possibilities* of statements by means of the mathematical concept of probability and the ψ-function.

... The mathematical inclusion, in quantum mechanics, of the *possibilities* of natural events has turned out to be a sufficiently wide framework to embrace the irrational *actuality* of the single event as well. It may also, as comprehending the rational and irrational aspects of an essentially paradoxical reality, be designated as a theory of becoming.

Since Pauli's text is often extremely compact, these passages need some explanation. The "possibilities" of events are described in quantum mechanics with the aid of the state function ψ. It is treated in a completely rational way: one gets the state function as a solution of the Schrödinger equation. The state function represents the symbolical picture of reality – the "abstract order of possibilities" – given by quantum mechanical formalism. The actual reality is represented by the registered results of observation. These individual experimental results are irrational (in the sense used here) because they are unpredictable: We can only give possibilities for the different results and their probabilities. This "irrational actuality" of individual experimental results forces us to abandon determinism, and it also makes it impossible to describe 'real events' in detail. We can only describe possibilities and give probabilities for different outcomes. Therefore the rational description is limited to these "possibilities"; in the individual events, i.e., in the detailed description of events, we meet something that exceeds the possibilities of human science.

It is important to point out, in addition, that in physics this situation is an implication of the introduction of Planck's constant. A deterministic description, i.e., an exact description of events in all details, seemed to be possible in (macrophysical) phenomena where one does not need to take into account the non-zero value of Planck's constant. In phenomena in which the inclusion of Planck's constant cannot be avoided, deterministic account is not possible. Simultaneously, realism in its traditional sense becomes impossible because it is not possible to define unambiguously the reality which should be the model for the theory.

The situation is, in fact, similar in all observations. We should always be critical with respect to the idea of realism when we must, in some respect, go far beyond the immediate sense perceptions. Even in astronomy, Lehti's speciality, one should take seriously the criticism with respect to ontology and realism presented in microphysics.

Some details in Pauli's quotation also deserve attention. Pauli points out that it is not possible to subdivide the phenomenon without essentially altering it. This refers to the indivisibility (wholeness) of a quantum phenomenon which Bohr repeatedly emphasized. In the new approaches to quantum theory which are called 'realistic', this feature of microphysics is neglected. This kind of 'realism' results in imagined theories without a sufficient basis in observed facts.

At the end of the quotation, Pauli mentions the "theory of becoming." The irrationality of reality implied by statistical causality makes it possible that the conception of reality, which normally is based on invariances, becomes generalized in the sense that it includes the possibility of the creation of something essentially new. Therefore, quantum theory is able to describe both being and becoming. This aspect of the theory has not been truly understood so far.

Lehti has repeatedly wondered what the indeterminism of nature and the statistical laws have to do with the psyche. Here we meet a question which Bohr was not willing to discuss but which cannot be disregarded if we wish to understand why 'independent reality' escapes us. According to d'Espagnat independent reality remains behind a veil which the nature of observations creates between the observer and the 'reality itself'. (*d'Espagnat* 1994. See also *d'Espagnat* 1993, p. 166.) This mystery is not, of course, characteristic only of microphysics. However, in physics this problem was ignored until the microphysical observations forced (some) physicists to analyze observations in this sense. In fact, even today very few physicists have made it clear to themselves that empirical science is associated with such limitations. In general, physicists like Lehti wish to maintain the realistic attitude in the traditional sense and don't like to admit that there are such limitations in the empirical method as some 'philosophizing' physicists – for example Pauli and d'Espagnat – have claimed.

I shall further try to illuminate the role of the psyche in observation, taking into account questions that Lehti has asked in this respect.

Lehti points out the fact that Pauli's ideas have not been discussed seriously in the literature. I think the reason is the reluctance of physicists to really study Pauli's ideas, which strongly deviate from the general patterns of thought in physics. Lehti suggests that "the reason can be that Pauli did not discuss questions of this kind in his articles published with his own name, while von Neumann, Wigner and some others presented their views of the influence of the psyche as clear public statements." (*Lehti* 1994, p. 101.) This is not true. Although Pauli was very critical with respect to his publications, he has published an exceptional number of philosophical essays also discussing the role of psyche more clearly than the other creators of quantum mechanics. His Kepler article (*Pauli* 1952), especially, is profound. I do not think that Lehti has taken enough time for reading Pauli's articles, because they are very compact and not easy to understand in detail.

In fact, Lehti has taken a stand with respect to Pauli's views in his book *Tanssi auringon ympäri* (Dance around the sun) where he describes thoroughly Kepler's scientific work. He mentions Pauli's Kepler article but his description of it is very superficial and erroneous (*Lehti* 1987, pp. 403–404). What else can one expect if Pauli's Kepler article is read on the basis of naive realism?

Lehti's remarks concerning the role of psyche in observations are slanted because he, like Selleri, seems to presuppose that this should mean causal effects on physical phenomena. It is not necessary to discuss these remarks in detail. Lehti writes, for example, that people seem to associate with observation "two strange phenomena from the point of view of the quantum mechanical formalism, at first the 'actualization' and then the 'reduction of the ψ-function'." (*Lehti* 1994, p. 103.) This, indeed, is the case, as was stated above. If one understands the Copenhagen interpretation correctly, these two 'strange phenomena' are really needed in the description of observation.

Lehti writes:

From the discussion concerning Wolfgang Pauli's philosophy, one can see that the novelty introduced by quantum mechanics is vindicated with the aid of some specialities of the mathematical formalism of quantum theory, such as the 'reduction of the state function'. We shall now begin to speak of 'observation' as a reality of this world, and then it is questionable whether there is any difference at all between quantum mechanics and classical physics on this point of practicing science. An observer does not dwell in the 'world of classical physics' or in the 'world of quantum physics'; he is living in this our common world, where nobody attempts to describe several phenomena, no more by using classical physics than by using quantum mechanics. (*Lehti* 1994, p. 107.)

The reduction of the state function is the very heart of the observational problems in quantum mechanics, and it deserves a more careful consideration than this kind of superficial playing with words. It is not just a 'speciality of the mathematical formalism'; a careful analysis of the observation process shows that it indicates a factor which has been totally ignored in modern natural science. It is the point where consciousness knocks at the door of the materialistic natural science.

As the basis of his epistemological and ontological views, Lehti presents the evolutionary epistemology of Konrad Lorenz. This epistemology presupposes the existence of an independent reality which we study as detached observers. This is exactly the basic philosophy in classical physics which makes it impossible to understand microphysical experiments. If we try to approach ontology and epistemology on this basis, we ignore all the basic problems of quantum mechanics.

We must remember that the progress of a quantum phenomenon cannot be experimentally investigated in detail; we can only symbolically describe what 'takes place' between the initial state and the final state in a quantum phenomenon. If we claim something about this process 'in between', we say something that, fundamentally, cannot be experimentally verified. Therefore it should be understandable that the picture of the microworld and of the microphenomena is not objective but that the human psyche has an influence on the shaping of this picture. Such are the consequences of the discontinuity introduced by the quantum of action. The result is that we cannot describe the microworld but only our knowledge of it.

d'Espagnat has called the 'picture' which science gives of the independent reality empirical reality. (See, e.g., *d'Espagnat* 1993, p. 56.) It is a picture of reality which changes in the course of time, the reliability of it being judged on the basis of the criteria applied in science. We cannot claim that empirical reality approaches the 'correct' picture of reality. Empirical reality is rational by definition because science must be rational, but implicitly it even receives influences from those aspects of reality which science is not able to describe: In observations which form the basis of empirical reality, the irrational aspects of reality are also effective. What kind of picture of reality is obtained in this situation obviously depends on the modes of action which are characteristic of the human psyche.

Characteristic of quantum mechanics are the 'matter waves'. A wave is associated with a great number of individual events taking place under the same experimental conditions. 'Wave' is a mathematical shape which is common for these individual events although the events themselves are independent of each other. Finding such shapes is characteristic of the human psyche. This is the origin of statistical causality: It is a shape created by the human psyche – but, of course, it must also have a counterpart in the outer world. "On the one hand a symbol is a product of human effort, on the other hand it is a sign of an objective order in the cosmos of which man is only a part," writes Pauli. (Letter to Fierz on 12.8.1948 in *Laurikainen* 1988, p. 21.)

It is characteristic of the human psyche that it finds some rationality in individual events which at first seem to be random. It describes what we know, on the basis of observations, of the phenomenon we are investigating.

I have devoted so much attention to Lehti's views because of his praiseworthy activity in this discussion. My response has been very critical, but probably Lehti represents a general attitude among the Finnish scientific community. I hope that this discussion will help people to see that the matter at issue is more profound than has been generally understood. *The fact is that a new, complementary conception of reality is forcing itself onto a par with the traditional ontological views, on the basis of convincing experimental evidence.*

B. The Criticism by Physicists

The attitude of my physics colleagues to my work can best be described as supercilious. In particular, the idea of the irrationality of reality has been found to be an offense against the scientific attitude in general; the definition of this concept seems to be less important. Because of such experiences, I have the feeling that science has in fact taken on the role of religion, trying to control the orthodoxy in dogmas and their publicity in the same way as the Church in Galileo's time.

I have not often had the opportunity to discuss with physicists the validity and importance of Pauli's philosophical views. My colleagues in general do not participate in public discussions about such matters.

In the rare cases of relevant interaction with some colleagues, I have obtained the following idea of their stand with respect to my attempts to strengthen the research and teaching of natural philosophy in the university:

1) Pauli's philosophy has only historical interest with respect to a certain phase of the development of atomic physics. The basic research in physics since the 1960s has shown that the difficulties which the founding fathers of quantum mechanics saw in this theory can be overcome and for the most part have already been cleared up.

2) The idea of the irrationality of reality cannot be accepted in science, which by its very nature is a rational endeavor. Ideas in this direction can only be understood as an attempt to fit religion into a scientific picture of the world, and such attempts are absolutely condemnable. "As far as we know, not even theologians accept any mixing of science and religion."

3) The introduction of psychic problems into physics cannot be accepted because it distracts attention towards useless deliberation. Physics has an excellent research method which continually produces good results. If psychic matters were mixed with physics, one would run up against such notorious questions as the mind-body problem. Psychology is, as a science, enormously behind physics, and it is not able to give anything to physics.

4) It is good, of course, to have an interest in philosophy, but it is not possible to reserve any time for philosophy in the regular physics courses; there are many important and difficult questions which cannot be included in the physics courses.

5) Funds reserved for physics education cannot be used for philosophy. Governmental power now presupposes that funds are used taking into account the needs of the society. Therefore, supporting interest in philosophy is not possible in physics.

When I have tried to point out that Pauli's ideas can influence the direction of physics research in the future and that they, in any case, are of a certain interest in the general cultural discussion which is needed now, the answer has at best been an understanding smile. When Stig Stenholm reviewed my book *Atomien tuolla puolen* (Beyond the Atom) in *Arkhimedes* (1A / 1986), he rejected the idea that Pauli's philosophy could have an influence on the direction of physics research. Here is a passage from this review:

It is not clear to me that 'a premature death cut short this maturation process which would surely have yielded fascinating and important results' in physics. Perhaps it would have led to a new understanding of the theoretical functioning of the psyche.

Stenholm also rejected the concept of quaternity – as far as it is not defined in an acceptable way. Thus, irrational matters first have to be made rational before one can speak seriously of them!

Pauli himself often spoke in his letters of a 'new science', in contrast to the traditional science. He felt it his obligation to speak of it publicly as well, but obviously because he was afraid of the criticism of his colleagues he remained silent. This caused him a difficult inner conflict and he interpreted certain, often recurring dreams as expressions of this conflict. (*Laurikainen* 1988, pp. 89-90. *Van Erkelens* 1993.) On the basis of my personal experiences, I understand Pauli completely – also with respect to the depression which can result from such a situation.

Stenholm's opinions have been among the most objective judgements I have heard from Finnish physicists. However, even he finds it difficult to accept the irrationality of reality – i.e., the possibility of aspects in physics which cannot be described by using the exact language of science. The end of the book review may illuminate Stenholm's attitude still further:

The book written by Laurikainen discusses interesting and thoughtful matters. The presentation of Pauli's letters to a wider audience is a valuable task, and the interpretation, in a popular way, of the ideas and concepts appearing in them is an interesting although difficult challenge. I find this book richly yielding reading for everybody who is interested in the problematics of modern physics. Its value could have been still higher if Professor Laurikainen had abstained from complaining that his ideas raise opposition. On page 117 one reads: 'The reaction of the academic world has surprisingly been strongly repelling'. This is certainly not an expression of an unconscious fear of the researchers of the unknown but rather a reaction to Laurikainen's attempt to present his writings as contributions to research in natural science. This is not the case even in this book but they are rather popular writings in the field of the history of science and ideas. As such they are welcome and they should be acceptable even for physicists. (*Stenholm* 1986, p. A 104.)

My view concerning the importance of philosophy for science is different. I would like to say that physicists, in their attitude towards Pauli's philosophical thought, have done something that should not be accepted without criticism. *Physicists take the role of an expert in matters which they have not even tried to understand and present propagandistic remarks which maintain erroneous and superficial opinions about truly deep ideas concerning the direction of research in physics today. Discussion of these matters must be allowed if scientific research is not just dogmatism but is interested in free criticism of its own basic trends.*

As a typical example of the style in which I have heard criticism of my opinions – and also of Pauli's ideas! –, some fragments will be quoted here from a lecture which Professor Matts Roos gave last spring in the seminar *Havainto tieteessä* (Observation in science). The lecture was entitled "Kvanttimekaaninen realismi ilman havaitsijaa" (Quantum-mechanical realism without an observer). I have done my best to edit the spoken Finnish text into another language. (The material of the seminar will be published in Finnish by the University Press, Helsinki.)

The irrationality is not a property inherent to quantum mechanics. Man is irrational. This irrationality belongs explicitly to him. On the other hand, I cannot understand how the word 'irrational' could be used for a quantum-mechanical system or a classical system. I find it to be out of the question. The problem is that if we investigate a system such that all of these three parts, quantum-mechanical system, measuring device and observer, are mixed, then it is clear that man's irrationality gets mingled onto the other systems. If one wishes to solve the quantum-mechanical problem, one should isolate the observer from the experiment.

Let me clarify this by an example.

Quantum mechanics must be able to explain phenomena where there is no observer. Of course, there is often an observer in a phenomenon; that I don't deny. But instead I claim that what we know in physics on the basis of natural laws must have a stronger force of explanation. The theory should also be able to explain phenomena where we have not been observers and where results were obtained a very long time ago. Professor Niiniluoto has said that the object of theory is the real world. What I am saying is the same thing.

Consider a situation with only two systems, a quantum-mechanical system and a classical registration. The oldest stars in this world have an age of some 14 billion years. The age of the Earth is 4.7 billion years, but man has walked on it at the most some 2 million years. And what is the age of philosophy? At the most 100 000 years. Thus, it is clear that the human irrationalism, if it has an influence on some experiments, can only have had an influence for 100 000 years. Everything that took place before it must be explained in some other way – in a rational way.

Thus, one has to cut off psychic problems when discussing observations - this is the simple recipe for how to solve the basic problems of quantum mechanics and the question of irrationality. There is also a simple way to solve the mystery of the appearance of waves in atomic events, which means the interference of state functions (or their superposition):

Schrödinger's cat is a good example. It is, as you know, a poor cat which is shut into a box where some slowly disintegrating radioactive material has also been put. When one atom disintegrates, it triggers a mechanism which kills the cat. Schrödinger, who is outside the box, does not know when this happens because it is a quantum-mechanical phenomenon: It depends on the moment when the first atom disintegrates. According to Schrödinger, this cat is not strictly alive but in a quantum-mechanical superposition of two states: 'dead' and 'alive'. Here a quantum-mechanical system is brought to the classical level. There has been much discussion about what Schrödinger in fact can observe or conclude in this experiment.

Rather, I would like to let Schrödinger go home and sit down myself near the cat. I would tell the story in the following way. There is in this room some radioactive material, and a terrorist has arranged that when the first atom disintegrates, a bomb will explode and we will all be killed. Then we are in the same situation as the cat. Can you now claim that your reality is impossible to describe? I have not changed anything else except that I have gone to the vicinity of the cat in the initial situation. The cat knows that it is quite alive but does not know what will happen. The future is always uncertain, and we cannot describe reality so completely that we can predict everything – including all quantum-mechanical measurements and what will happen in them.

I think that much confusion is caused by an idealization. ... We really do not recognize that we are sitting here as a linear combination of two states, both living and dead at the same time.

A small quantum-mechanical example which I often use in my teaching is as follows: Here is the floor plan of a room with two doors. The observer sits in this room and its two doors are closed. In the surrounding larger room a cat and a dog are moving freely. Now the observer hears scratching at both closed doors: an animal wishes to come in from each of them, but he does not know which animal is at which door. Classically it is either a cat or a dog but quantum-mechanically both animals are equal, they are both cat-dogs: they are linear combinations of the cat state and the dog state. When the observer opens one door, an Einstein-Podolsky-Rosen phenomenon takes place: the cat-dog state is suddenly transformed into a cat. And simultaneously the animal behind the other door is transformed into a dog, without even knowing that the first door was open.

At last, Professor Roos finished his description concerning the problems of quantum mechanics as follows:

I have some proposals for the solution of these problems. First, this wave-particle dualism is no problem today. The waves are not physical but probability waves. Really, more than 70 years have passed from the birth of quantum mechanics, and we are quite familiar with probability waves.

Then the double-slit experiment is a test for this property but not at all for other problems of quantum mechanics. I find it a very weak example.

Further, the fact that quantum mechanics gives just probability predictions does not disturb me. I would like to say that it is of the same kind as the story of the shaggy dog. Don't you know it? A lady in New York went to see her veterinarian and said: "What shall I do, this dog is so shaggy?" The veterinarian tries to do something but nothing happens. Then this lady goes to London, tries the same with another veterinarian, but the result is the same. The same happens in Singapore, aso. At last she comes back to New York and goes to see another veterinarian. This looks at the dog and says: "It's not so shaggy." This was the story. So if the theory makes only probability predictions, it appeared quite shaggy 70 years ago, but no longer.

Finally, the sudden reduction of the state. The classical device disturbs the system and its state is reduced because of this. However, if you find this unsatisfactory, remember that this is a statistically causal phenomenon. Somehow we don't accept it according to the common sense and wish to have a deterministic explanation. That is not possible. Thus, the problem arises because we don't accept statistical causality. Einstein did not accept it.

So most of the problems in quantum mechanics have been cleared up. Oh yes, then that problem concerning reality. From the time perspective of Schrödinger's cat we learned that if the state before the measurement cannot be defined this implies that the state does not include knowledge of future measurements. If we think ourselves into that state before the measurement, so we do not know what kind of measurement will take place and therefore we cannot have complete information. A corollary of this is: no hidden variable theory can include knowledge of all future measurements. Therefore the concept of reality cannot have any meaning for a physicist. It is unclear in a way, but I think it is unclear in the same way as the future is unclear.

The conclusion is that one has to choose the problems correctly. Then one can show that quantum mechanics is a theory which has to fulfil certain physical requirements and not psychic requirements. Thank you.

This lecture was a typical presentation of the problems of quantum mechanics given by a physicist, even in the sense that philosophical problems are understood to mean mainly the popularization of physics. I think that this describes in general the attitude of physicists towards philosophy because in a philosophical discussion one speaks of matters which in physics are described in the elementary courses only.

Some critical remarks are perhaps necessary with respect to Matts Roos' lecture from the point of view of philosophy.

In the first quotation, Roos finds it impossible to speak of the irrationality of a quantum system. He finds it to be characteristic of the human level, of the properties of the human psyche. He polemizes against the use of the word 'irrational' in a sense other than he is accustomed to using it. One could expect that respect for Pauli would presuppose that one first looks at what Pauli means with a certain word, especially as Pauli's articles are extremely exact. Roos would like to cut out the observer when speaking of quantum mechanics, but then one cuts out precisely the problems concerning observations which we are discussing here. Then there is naturally no need to speak of 'soul' or of 'consciousness' in the observational problems! So simple is the elimination of philosophical problems for a physicist.

Schrödinger's cat is not a good example of the superposition of states in quantum mechanics because it is macrophysical. However, if we understand the state functions and their superposition (interference) to be a description of 'our knowledge', then this famous example is illuminating. That a cat at first is a superposition of a 'living' state and a 'dead' state only means that we don't know, after the poor cat was shut into the box, whether it is alive or dead at a given moment. On the basis of the half-life of the radioactive material, we can estimate, however, what the probability for each case is at a given moment.

But this interpretation is not acceptable for materialistic realism because one needs 'knowledge' and thus 'consciousness' in the description of the world. When Roos says that he considers the theory to concern the 'real world', he probably means the 'world of matter' measurable with the aid of physical devices. From his argumentation, I cannot see what the nature of the state function is in his 'real world'. It seems as if he is speaking of a 'cat state', of a 'dog state', and of a 'cat-dog state' in some kind of realistic sense, but then his illustrative example of the EPR phenomenon with two animals is truly strange: then the animal at door B is really transformed into another being when the observer looks through the open door and sees that the animal scratching at door A is a cat. Probably the critical gaze of the observer also changes the strange cat-dog animal at door A into a definite cat! How these 'real' changes are possible Roos does not tell us.

If one is allowed to consider the state function as a description of our knowledge, such strange transformations become unnecessary – but then the consciousness begins to haunt us again: The changes do not happen in the physical outer world but in the consciousness of the observer. The situation is described thus in the Copenhagen interpretation, but unfortunately it is antiquated!

The astronomical example of Roos is also strange. I have seen that philosophers wonder about the fact that Bohr always speaks only about those quantum effects that are observations, yet quantum mechanics also has to be applied to astronomical phenomena which took place in the immemorial past for example. Recently, even some physicists have begun to wonder about such questions. However, it was not Bohr's idea that quantum mechanics would only be applied to quantum phenomena where there is an observer. All the time quantum mechanics has been applied to questions concerning the structure of matter, for instance, where it is not possible to think that there is an observer present at each quantum step! Bohr carefully examined observations only because our physical knowledge is based on such phenomena.

Perhaps this strange idea that the Copenhagen interpretation would not allow the application of the theory to phenomena other than observations is caused by the criticism with respect to the concept of independent reality on the basis of the original Copenhagen philosophy. Then the picture of the universe, for example, appears to be only a theoretical construction. So I indeed understand the basic philosophy of the Copenhagen interpretation. The realistic belief of Roos in the history of the universe according to the present theories I cannot share. It is necessary to be critical with respect to all extrapolations which go very far from the field of our immediate sense perceptions.

First of all, we always have to remember the lesson obtained from the observation that *quantum processes take place step-wise* and that in each 'quantum jump' there is inherent an *irrational choice* between the possible final states. I find the application of mathematical theories to the origin of the universe and to its initial development to be a clear mistake. Quantum-mechanical ideas are applied there in a field which is too far from the situations which form the empirical basis of the theory. At least one should then be exceptionally careful in not deviating from the background philosophy which was the basis of the formalism. In quantum mechanics this background philosophy contained implicitly an aspect of spirituality or, in other words, an aspect of creation. One should not bury this aspect under patterns of thought inherited from materialism – especially when making an attempt to describe creation. This cannot be described in a mathematical language. The origin of existence is in an irrational dimness. If one wishes to say something about it, the language of a poem is more proper than the language of a mathematical formalism. For example:

Who then knows, who has declared it here,
from whence was born this creation? The
Gods came later than this creation, who
then knows whence it arose?

He from whom this creation arose, whether
he made it or did not make it, the highest
seer on the highest heaven, he forsooth
knows, or does even he not know?

Hymn of the Rigveda, X, 129. (According to S. Radhakrishnan: *Indian Philosophy*, Vol 1, p. 100. See *Rosenfeld* 1979, p. 463.)

Atoms and Biology

An example of the dismemberment in science today is the attitude of biologists to the philosophy related to quantum mechanics. Biology is now developing strongly and has very promising prospects for the future. It is symptomatic that biology has obtained contact with geometry in the form of the double helix of the DNA.

Understanding geometry was the entrance requirement to the first Academy. Now biology begins to fulfill the characteristics of science!

The analysis of the concept of causality which has been necessary in atomic theory should have implications for biology. Free choices, i.e., a purposeful reaction to the requirements of the circumstances, are characteristic of living organisms. The possibility of teleology which statistical causality opens seems to correspond to the most typical aspects of life. However, biologists seem to be still more bound to determinism and materialism than physicists are.

I am not the only one who has had this experience. Let us quote a passage from d'Espagnat's book *A la recherche du reél* (In Search of Reality) entitled "Objectivity" (*d'Espagnat* 1983a, pp. 51-54; see also the German translation *d'Espagnat* 1983b):

Atomistic, or mechanistic, materialism is a set of two assumptions. An ontological one: Independent of us, the world is made like a clock. Tiny bodies, fields, and forces constitute its parts and its springs. An epistemological assumption: We are all able to know better and better, and finally quite well, the world just as it is, with all its composite wheelworks. Of course, these propositions cannot be *derived* from any obvious truth. Under the incentive of various views, both philosophers and mystics can therefore quite well discard them without inspection. But their upholder (who, at present, is, as a rule, the biologist) then has a ready-made reply: A conception, he will answer, can only be justified by considering its consequences. Now what are those of the philosopher's standpoint? Or of the mystical standpoint? Objectively, there are none whatsoever. Whereas, look at my own: it is verified by the totality of classical physics, by quite large part of astrophysics, and by the whole of contemporary biology. Even in fields such as those of life and thought, in which, naively, we could have believed in finality and in an active role of mind, our present day discoveries show the reign of pure chance and necessity. Think of the model of the twin helix!

Thus, this scientist proceeds, would it not be foolish or, at any rate, infantile to go on talking about final causes? Or, again, foolish to consider mind (and the consciousness we may have of things) to be *entities* on the same level as matter, that is, on the same level as the tiny bodies or fields which, taken alone, account for the whole universe including man and including man's mind and consciousness? These are all childish views, or, better to say, secondary effects, bewitching and luring semblances from which every strong lucid adult should free himself. A multitude of tiny bodies bound together by forces described by quantum physics, sometime subject to determinism and sometime to objective chance – such indeed is the ultimate stuff of the world. Everything comes down to physics, to the pure and icy objectivity of physics!

Overawed by such a triumphalism and such a bulky self-reliance, idealists and thinkers have to hang their heads. They all confess that they, up until now, gave themselves up to trifles instead of beholding the very essence of mankind, which is, of course, molecular biology. But, by the way, is it actually molecular biology? No indeed. If it is true that (as our scientist was just saying) biology reduces to physics, at least in principle, then physics, not biology is at the heart of the matter. The most inquiring minds within their troup (but these are not many, this track being more uphill than the former!) therefore turn to the physicists, to learn what *they* have to say on this point.

These sing another tune, and their triumphalism is different. Less youthful to be sure, but no less powerful. For some thirty years now, basic physics, laboring in the field of artificially produced phenomena, has been digesting – painfully at times! – its pre- World War II conquests. Indeed, what does the physicist assert? Just one short simple but true proposition: "In principle, I can account for the totality of the phenomena that you see around you – All of them really!" We reply with astonishment, "Yes indeed, really all – How then? – Basically by means of the Maxwell and Schrödinger equations."

At this point, we divines and thinkers are overawed even more. "This indeed corroborates what our mechanistic interlocutor was saying a moment ago!" we exclaim together. But "Hush!" the physicist whispers, "the word *mechanist* is not in our good books here." "But why?" we ask. "Don't you agree with, say, the biologist, that science, and science alone, is objective?" "Yes, to be sure," our man exclaims with determination. "But, then, why do you brush aside this qualification of *mechanist*?" "Oh, well, the point is just that these biologists and us do not agree entirely on the meaning of the word *objective,* so that the term *mechanist* somewhat offends our ears. However, please believe me, this is but a minor irrelevant detail; its explication would just bore you." "But, look here, we must know what we divines and thinkers should tell our flock. What should we say to them about the nature of science?" "Well, just that it is objective. Isn't that extremely easy?"

These words bring about a general hush that the physicist takes to be skepticism. To try to convince us, he stresses the point: "You see, he says, between religious orders – and conceptually the scientists are such ascetics that, in a way, they are our modern monks! – between religious orders, as I was saying, this is not the first time that such – oh so minor! – difficulties have arisen. But, drawing our inspiration from what the shrewd author of the *Provinciales*[1] reports, we and these biologists did in fact succeed in making peace on this subject. For that, we just had to agree to say in unison that the basic principles on which science is based *are objective ones*, full stop. As a matter of fact, we, the physicists, do not understand that statement as most people do: in our interpretation, the principles in question *can refer in an essential way to the abilities or the inabilities of the observers,* provided they should be common to *all* potential human observers instead of being particular to one of them. Indeed, we are tied down to such careful formulation; otherwise, conventional atomic physics would be inconsistent. Crumbling down, it would sweep away molecular physics in its fall, and molecular biology as well! On the contrary, when biologists say that a given statement is objective, the idea they are trying to convey is that it bears on Reality itself, of which men are merely an accident, so that it should obviously not

---

[1] Blaise Pascal

refer in an essential way to them. *That* is why we, quite appropriately, call them *mechanists*, which is among us a disparaging epithet. But these differences remain secret, for, as you may well grasp, it would be unseemly to display such discordances! The laymen would not understand! Thus, to those who ask, you should just answer that science is *objective*, and you must be very careful not to try to define that word."

„But, Reverend Father – oh, sorry, we meant Professor – we then reply, still, this *does* make some difference. For after all, if it is true that the basic principles of physics cannot even be formulated without some essential reference to the abilities of human observers, or even merely to the limitations of their general faculties, then is it not sheer nonsense to go on speaking of *naked apes* and all the other leitmotivs of a materialist world view which fancies itself to be in the vanguard? Man, then, would *not* just be a negligible physical system. He would *not* come out of Nature, as a mere adventitious petty excrescence, in but a tiny portion of the Universe that his science describes and that his senses grasp. On the contrary, he would be the very measure – and even, finally, the co-author – of all this empirical World that he perceives and that he believes exists all by itself. Protagoras, not Lucretius, would have been right!"

„Please excuse me," the scientist replies, "specialization is binding: I simply may not listen to nonscientific remarks!"

This nice description of the situation in biology and in physics, written with the French sharpness, is quoted on the kind permission of Prof. d'Espagnat and Editions Dunod. To the biologists I would like to state with emphasis that *teleology is no longer a nonscientific idea*. It is difficult to understand how biology can ignore the idea of teleology. This has happened under the pressure of deterministic causality inherited from physics, but physics is not deterministic any more.

If we accept the idea that teleology is the very characteristic of life, the controversy between the biological theory of evolution and the idea of creation disappears: *evolution can be understood as continuous creation*. Some of the creators of quantum mechanics already pointed this out 70 years ago. Biologists who quite generally still support the deterministic picture of the world have so far rejected this idea. One should realize, however, that at least on the level of microphysics – in phenomena where Planck's constant plays an essential role – determinism has proven to be a Utopian idea. For example, changes in the structure of DNA take place on the level of atomic bonds. Determinism certainly has its limits, and beyond this limit causality can be only statistical.

Simultaneously with the discontinuity introduced by the non-zero value of quantum of action, the irrationality of reality has to be accepted (in the sense the term 'irrational' is used here!). It means that in the world there are aspects which cannot be described by the methods of present science. Reality is not in all of its aspects measurable with physical devices and methods.

One should think seriously of the possibility that life does not mean only physical and chemical processes. The strengthening of materialism implied by the deterministic conception of causality during the last three hundred years is a great mistake. It is time to get rid of it – especially in the phenomena of life. *Perhaps teleology is the very characteristic of the idea of life:* a freedom in choices and a purposeful meaning in them, against the teachings of materialistic science. The mystery of creation cannot be solved by the methods of the rational science.

III
Creation

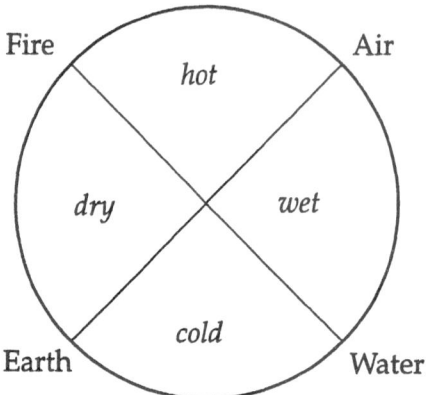

From the trinity to the quaternity.

The hypostases of the Godhead and the problem of evil. *Above*: A figure visualizing the four elements (the 'hypostases' of matter). *Below*: A figure by Scotus Eriugena visualizing the hypostases of the Godhead (1. God the Father, 2. God the Son, 3. the Holy Ghost); he interpreted the fourth hypostasis as deification, returning to God (the trinitarian thought). If the fourth hypostasis as well is interpreted as a person, it represents the Devil. This opens the possibility for real changes in the world – for continuous creation (the quaternarian thought).

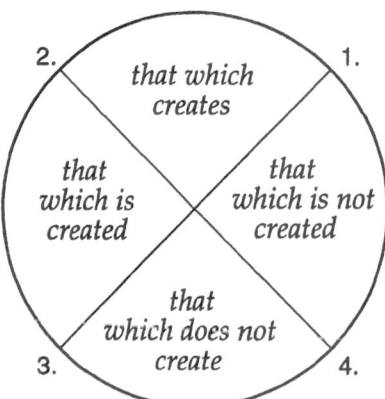

9. Freedom

Skeptic: I have tried to penetrate into your ideas concerning irrationality as an aspect of reality and the limits of science, but I must say that they have smack of subjectivity. You have said that Christianity is your faith, and it looks as if the truths you present would get their color from this fact. With the aid of irrationality you find a place for the supernatural in your picture of the world. Don't you try to explain, in a propagandistic way, your fawning on religion while, in fact, you in your old age have deceived the sane scientific attitude? I don't really like this.

Physicist: The matter is often understood thus, it is true. The ruling dogma in our country is that science and religion refer to different spheres of existence which have nothing to do with each other. For most people, it seems to be impossible to believe that concentrating on the actual problems of natural philosophy has made religion alive to me. As I understand it, the witness of nature is now needed in order to break the repression of religion which is enormously strong in many people. Educated people, especially, have become estranged from religion because this is expressed in such a form that it seems to be in contradiction to the scientific conception of truth.

S: One can have different opinions about that. Christianity is already outdated, and it is only natural that people increasingly base their views of life on a safer foundation than on fancies about the resurrection of the dead and paradise as a reward for the sufferings in this world.

You have declared, haven't you, that the irrationality of reality means that there are matters which cannot be investigated by scientific methods at all. You have specifically said that certain questions cannot be asked by using exact concepts but that some dim language is needed which you call *episteme*. It is natural that people who are accustomed to the scientific usage of language cannot accept such remarks. The progress of science is based on the exact definition of concepts and on the logically clear formulation of statements. If one accepts compromises in these respects, then one can present all kinds of stories as true.

P: I have entered into extensive correspondence about such things; my correspondence with Henry P. Stapp, in particular, has been very illuminating. In recent years, he has been very active in the investigation of the basic philosophy of quantum theory, and recently he published a book about consciousness on the basis of quantum theory (*Stapp* 1993). He has attended the symposia on the foundations of modern physics in Finland – twice in Joensuu and in 1992 in Helsinki. After the symposium in 1990, I corresponded frequently with him concerning the irrationality of reality, i.e., the question of whether quantum theory forces us to accept the existence of matters which must be called real but which cannot be described with the aid of rational theories.

Stapp's quantum mechanical theory of consciousness is based on the idea that consciousness and the physical state of the brain are isomorphic, i.e., they have the same structure. When I stated that I cannot consider this connection so simple, Stapp wrote that with his theory he only wishes to put the idea of psychophysical parallelism of Spinoza, Leibniz, and others into an exact form: The world of matter and the world of spirit are parallel, and both express the same features.

S: Well, does not his theory use a fully rational language, perhaps even mathematics, when describing the phenomena of consciousness? Don't you think, instead, that the 'spiritual' matters – or 'psychic' as you wish to call them – are just such matters where an exact scientific description is not possible?

P: I have, in fact, written to him that I cannot consider that the description of consciousness will succeed in so simple a way. I find a strong attraction to the idea which Pauli presented in several connections, that we must abandon Cartesian dualism and consider 'spirit' and 'matter' – or as I wish to say: the 'psychic' and the 'physical' aspect of the world – complementary to each other. Thus, we try to describe these matters by using the concept of complementarity introduced in quantum mechanics. The 'psychic' and the 'physical' are inseparably associated with each other, that's true, but we cannot presuppose that their structure is the same. They are complementary in an analogous sense to certain concepts in quantum mechanics that are complementary to each other, and their mutual relation is a very profound problem which cannot be solved by using the methods of the present science. Pauli found the illumination of the relation between 'psyche' and 'matter' and the search for a new conception of reality to be the most important problem of our time.

S: Here we again meet Pauli, whom you admire so much. Are you not bound excessively to his views – especially since it seems to me that other physicists do not rate his views as highly as you do?

P: I regard Pauli very highly as a philosopher, that's true, and I have learned more from him than from anyone else – partly for the simple reason that I have been engrossed so long in his thought. I find his philosophy a revolutionary pattern of thought which is sorely needed in our time. Among the physicists in the

Copenhagen school, he most clearly disengaged himself from the positivistic philosophy, and he understood the philosophical implications of quantum mechanics more clearly than did any other physicist.

Among the physicists, Pauli is esteemed on the basis of his achievements in physics proper. The reason why his philosophical ideas are not rated as highly as I rate them is that people do not know and understand them yet. The only more-or-less uniform presentation of Pauli's philosophy is in my book *Beyond the Atom*, and his philosophical essays have only recently been published in English. There are also features in Pauli's philosophical thought which are in strict disagreement with the dominant philosophies of today; for physicists, they are very strange, and their publication has not been easy. I have the impression that very influential physicists find the philosophical ideas of Pauli to belong to his 'dark side', and obviously some people had hoped that they would be forgotten.

S: It is often difficult to know whether you refer to Pauli or present your own ideas.

P: No doubt, that is true. Perhaps it is best to say that everything that I present in this discussion must be ascribed to my account, especially all errors and mistakes. Once more, however, I acknowledge my enormous debt of gratitude to Pauli. Without penetrating into his letters, it would not have been possible for me to achieve such a general view of the nature of human knowledge and of the fundamental problems of existence as I have succeeded in forming since my retirement.

Everything cannot be found in Pauli's letters, of course, but they form an extremely rich starting point, and they have forced me to study more carefully than before the views of many thinkers and also to become familiar with many fields and matters which had been virtually unknown to me.

Thus, I believe now that I understand Pauli's thought quite well, and in general I have fully appropriated his views. It is possible, of course, that I have misunderstood him in certain matters, all the same. Therefore, I do not claim that in the following I present Pauli's philosophy; I describe such views which I have learned when I have tried to re-orientate myself in philosophy on the basis of Pauli's thought.

S: These remarks concerning Pauli's philosophy were caused by your referring to his views about the psychophysical problem. You obviously think that in this particular case we need the obscure 'language' which you call *episteme*.

P: I must say so, because this is a typical metaphysical question. It seems to me that such concepts as 'consciousness' and 'soul' are often considered completely unscientific. It is not possible to define them, at least not with the accuracy which is required in the exact natural sciences or in analytical philosophy. Many people therefore avoid their use. I find that atomic theory clearly leads us to the view that there is in reality an element which is of a quite different nature than the 'matter' or 'energy' which are studied in physics. We must dare to speak of this element of reality although we cannot precisely define the concepts which are needed when describing it. We come here into the realm of metaphysics; in particular, metaphysics which refers to these kinds of problems must now be approached in some way. Here one needs ways of description which do not fulfill the requirements of strict rationality.

I find it incorrect to say that such attempts are 'nonscientific'. We have, in fact, some intuitive insights into the need for the concept of 'soul' in addition to matter; it is by its nature something totally different from matter but its reality, however, is obvious. We never have the possibility to make the description of our experiences belonging to this sphere more exact if we do not dare to speak of matters which cannot be described in the way required by strict rationality.

S: Your desire to speak of 'soul' is precisely one thing which strongly reflects a religious basic attitude. I hope that you don't by the same token claim that the soul is immortal? I think you should in some way define the 'soul' if you wish to speak of it in the scientific sense.

P: Let us avoid the question of the immortality of the soul today. Instead, I am willing to discuss the question of what I mean by 'soul'. My starting point is atomic physics because there we have met very exact matters which, as far as I understand, force us to speak at least of the consciousness. This is a concept which many people find quite necessary to introduce into the conception of reality which is shaped on the basis of atomic physics. The so-called paradoxes of quantum mechanics disappear if we are ready to say that the state function describes 'our knowledge' and that in an observation our knowledge about the situation is changed. Therefore the state function is reduced in an observation in a way which cannot be theoretically governed: In an observation, nature gives us information and we 'become conscious' of it. Indirectly this presupposes that we acknowledge the consciousness as an element of our picture of the world; it is not described by quantum theory, but it is necessary to speak of it if we wish to offer a view of the basic nature of the quantum world – of ontological questions.

S: Do physicists in general think that it is necessary to include in reality an element of this kind, let us say a 'spiritual' element?

P: No, physicists in general do not discuss ontological questions, and in general they do not find the inclusion of consciousness necessary. There is, in fact, a small group of physicists who try to approach the basic problems of quantum theory according to the rules stipulated by philosophers, which means in the manner of analytical philosophy. This group has proposed a lot of different approaches which presuppose a materialistic conception of reality. It is important to realize that not one of these new approaches, which then compete with the Copenhagen interpretation, has been entirely satisfactory; they always have problems which usually concern in some way the nature of an observation or measurement. None of them has been generally accepted, and the majority of physicists are not especially interested in these attempts which are philosophical by nature.

I take the original Copenhagen interpretation for the starting point of my considerations; it has been generally accepted by physicists, and in applications, when theory is compared with the measurement results, physicists in general use its pattern of thought. This interpretation works well on the level of practical applications. I do not see any need for other interpretations: This original interpretation belongs inseparably to quantum mechanics, as I see the situation. A mere formalism, i.e., a collection of formulas, is no physical theory; the theory must always be interpreted, and in quantum mechanics the natural interpretation is the Copenhagen interpretation.

S: Thus, only very few physicists have the opinion that quantum mechanics 'forces us to speak of the soul', isn't that so?

P: That's right, but the situation is very strange in that physicists simply refuse to speak of the philosophical foundations of their science. Therefore, the philosophical problems of the Copenhagen interpretation have remained unclear. Philosophers and the group of physicists mentioned above have, in fact, discussed them extensively, but in general they present such high formal requirements that the empirical lesson of quantum mechanics has become muddled by logical considerations.

I think that Pauli's philosophy contains the clearest view about the ontological implications of quantum mechanics. I hope that this view, which so far has been ignored, will now be seriously discussed.

Thus, I propose that we begin our scrutiny from the Copenhagen interpretation and from the view that there is, along with matter, another basic and fundamentally different element in reality which appears in the form of consciousness. However, the unconscious functioning of the human psyche must simultaneously be taken into consideration – i.e., the functioning which precedes our 'becoming conscious' of something. I would like to call this whole which contains both the consciousness and the unconscious the *soul*. [Let us state, for the sake of clarity, that the word 'consciousness' corresponds to 'das Bewusstsein' in Pauli's German essays, and 'the unconscious' to 'das Unbewusste'.]

Of course, the soul also has other expressions which are closely related to its conscious and unconsious functioning. Let us especially think of the expression of spiritual life which is called will. It is associated with problems which in my writing I have described as expressions of the irrationality of reality. They are related to the same matters which force me to speak of consciousness and of soul when I try to clarify the conception of reality presupposed by atomic theory.

Today I would especially like to discuss problems concerning will and the question of how these problems are associated with consciousness and, a fortiori, with the picture of reality. In these matters, a great conceptual confusion dominates, and the lesson given by quantum mechanics can perhaps help to clarify it. We must remember, however, that we can approach these questions only in the sense of a science which uses the language episteme – as distinguished from a strictly rational knowledge, which Plato characterizes with the term *dianoia*. We are in the realm of metaphysics here.

S: With respect to metaphysics, I am very suspicious. But tell me, what kinds of problems you find when discussing the concept of will.

The Problem of Free Will

P: Will is the second main dimension of the spiritual life, in addition to rationality (reason). Will and reason are not independent of each other, but we shall come back to this question a little later. Reason is an analyzing, shaping, spiritual capacity searching for invariances, as if it looks backward in time to that which is safe, *which already exists*. Will has the future in view; it *creates something new*.

Especially in modern times, after natural science had appropriated a completely deterministic picture of the world, the problem of free will has become associated with the question: How is free will possible in a world which is governed by absolute natural laws? The understanding of spiritual life has indeed been very difficult on this basis, and therefore modern man is inclined to totally forget that he has a soul, which is completely different from the body and from the material world investigated in the natural sciences. The soul is close to becoming choked by the superabundance of knowledge concerning matter.

Now more and more people seem to realize that there is something wrong in this 'modern' pattern of thought. The soul is rising in protest. People and society have begun to feel unwell. The real problem of free will is related to such matters.

It is a fact beyond dispute that people have the capacity to make free choices between different alternatives. Morals are based on whether man has in a certain case acted right and in an acceptable way or wrong and perhaps in a way which demands punishment. People have a very great need to increase this freedom, so that one need not work like a slave at tasks which are imposed from the outside, but that one can, as much as possible, decide for oneself what one wishes to do. All of us meet restrictions which society imposes on us in this respect. How our life shapes itself depends very much on the freedom that we succeed in maintaining within the restrictions which circumstances and society stipulate. "Fate draws the willing, drags the recusant," said the old Stoics.

The freedom of will is one of the most obvious and most important facts which we meet in life. However, for philosophers it has been a great problem, especially in modern times. The choices of will are choices between different possibilities in certain situations, and each choice should have visible effects in the events which follow the moment of choice. But how is such a choice possible, if phenomena in the world are predetermined – if the world is a machine which performs according to absolute, deterministic laws?

Kant solved the problem by dividing the world into two parts: a *phenomenal world*, which is the object of the empirical sciences, and a *noumenal world* or the world of the 'things-in-themselves', which is beyond the reach of empirical science. The spiritual life belongs to the noumenal world; ethical questions, values, and religion belong there. The phenomenal world is governed by absolute causality, determinism, while the noumenal world is a world of complete freedom. Causality does not belong to the noumenal world at all. There, completely free choices are possible, and these should be based on the conception of good and evil which the conscience gives us.

In a way, Kant solves the problem concerning the freedom of will in a deterministic world: this freedom does not belong at all to the world of matter but to the inner world of man. There the categorical imperative of Kant plays the role of causality: "Act only on that maxim which you can at the same time will to become a universal law." But how is it possible that a free choice of human will can influence the occurrences in the material world if the world is a deterministic machine?

It does not, say the Kantians. The freedom of will is just an inner experience. We must try to adapt ourselves to God's will which becomes true, anyhow – as the old Stoics wished to say in their own way.

S: But is not the freedom of will something more than just an inner experience? People do not fight for freedom only for the sake of the peace of their souls! There is a question of certain very concrete solutions concerning the material world. A slave cannot become rich and fulfill his needs and his possibilities to the same extent as a free man. By our choices we try to increase our own freedom, and this presupposes the possibility of influencing the events in the material world.

P: This is exactly the case: the strict distinction between the phenomenal and the noumenal world is practically impossible. Certainly, by the decisions of will we can influence phenomena in the material world, i.e., in the phenomenal world which is the object of the empirical sciences.

It is to be deplored that Kant's and Descartes' patterns of thought still have a strong influence in the Western countries. This has given full rights to materialism within the natural sciences and created an unbridgeable abyss between the natural sciences and the humanities. It also helped theologians to fortify themselves in their own 'world of spirit' without noting that God's influence also appears in nature.

The reason for the strict Kantian distinction was in the deterministic picture of the world which he had appropriated from physics. Because spiritual life could not be included in the deterministic material world, it needed a space of existence of its own, free from determinism.

Now the situation is quite different as determinism has been abandoned in physics. There is reason to think that the idea of deterministic causality has to be completely abandoned (in other fields besides physics as well). Then there is no need to draw a strict distinction between the phenomenal and the noumenal world. Again there is a place for the soul in the world view of the natural sciences. I have justified this on the basis of the irrationality included in the conception of reality, which changes the basis of the world view in a profound way. Here we can examine this from the point of view of the freedom of will.

S: How can you justify the idea that the irrationality of reality will make room for the soul in the world view of the natural sciences?

P: Remember, please, that the interpretation of quantum mechanics explicitly presupposes the inclusion of consciousness in the conception of reality. I would like to remind you of complementarity, which Bohr emphasized in the interpretation; it states the experimental fact that the properties of a microsystem depend on the method of observation. Thus, the choice of the method of observation made by the observer – which presupposes a certain 'act of will' – belongs, as an essential element, in the reality of microphysics. We cannot speak of the microworld without simultaneously speaking of the observer, of his will (his choices), and of his consciousness.

If we do not impose materialistic or rationalistic preconditions for our conception of reality we must admit, I think, that besides the material world there is something which must be called real and which makes it possible for us to give shape to the world. This element of reality which gives shape to the stimuli we receive from the outer world is called consciousness, and it is an important expression of spiritual life. 'Soul' not only has a place in the picture of the world opened by atomic research, but according to my view, we are forced to accept it as an element of this picture because only then does the interpretation of the formalism become simple and natural. Attempts to avoid speaking of consciousness lead to complicated conceptions of reality, and in addition, none of these new approaches has been satisfactory in all respects.

S: I am inclined, at least, to take into account this possibility. Perhaps science is about to rediscover the soul. But how do you associate the freedom of will with this pattern of thought?

P: It is the most simple thing in the world. The world of quantum mechanics is not deterministic but governed by statistical causality. Thus, the laws do not determine the course of phenomena unambiguously but allow different possibilities; in each individual event, it is as if nature makes a choice between these possibilities. One can say that nature makes free choices all the time because laws of nature do not govern these 'choices'.

S: Obviously, laws in the atomic world are statistical by nature, but is it possible to draw conclusions from this with respect to the human spiritual life – for instance, the freedom of will? These are matters on two different levels: on the one hand, microphysics; on the other hand, happenings on the macrophysical level. In addition, statistical laws refer to the material world, and the freedom of will belongs to the spiritual reality which, as you have said yourself, has a quite different nature.

P: People continually present such objections to me. They no longer seem to understand that there is only one reality, and we must try to comprehend its basic structure. Even philosophers have adjusted themselves to the specialization tendency of Western culture, because the formal requirements of analytical philosophy prevent a discussion of the most general aspects of reality. People think carefully of details only and do not see the whole.

In philosophy, people speak much about the freedom of will, it is true, but the central problem then is whether free choices are possible in a world governed by deterministic laws – and what 'freedom' in this case can mean. If determinism has now been abandoned, the problem has to be seen in a new light. In fact, the problem has disappeared!

However, in general this kind of idea seems to be in some way forbidden for philosophers – and for many physicists as well. Full freedom, not bound by any laws – i.e., something by nature 'irrational', in the sense we here use this term – is so strange an idea and in contradiction to Western rationalism, that it cannot be taken as a real possibility. The respected Finnish philosopher G.H. v.Wright once said at a symposium that one must find some other manner of description because 'freedom' and 'will' are not concepts which can be associated with physical matters.

S: Such objections are quite understandable, because you do not mean, I suppose, that elementary particles have 'will' or that they 'make choices', although there are questions concerning individual events which cannot be answered by scientific theories.

P: This is, as far as I understand it, the basic problem of the Copenhagen interpretation. Exactly on this point Einstein and Bohr had different opinions. Einstein represented the rationalism of classical physics, and he saw in the indeterminism which statistical laws presuppose something forbidden, something not acceptable according to his scientific conscience. When Bohr claimed that the description of phenomena given in quantum mechanics was *complete*, this meant that reality is not deterministic but that we must accept a new concept of causality. Then there remain 'openings' in the description of individual events which cannot be filled in any rational way. Bohr refused to think what this means in the ontological sense. Pauli has clearly stated that it leads to a new conception of reality which contains both rational and irrational elements. Simultaneously, it led him to a philosophy of one world, i.e., to a conception of reality which contains both 'physical' and 'psychic' aspects.

When people claim that matters of different natures are mixed here, they do not understand that we are dealing with a new conception of reality where 'physical' and 'psychic' cannot be separated from each other. The objection you made is natural and strikes the heart of the question, but it must absolutely be rejected if we wish to discuss the philosophy of one world which I would now like to illuminate.

S: But what about the remark that one should not draw conclusions from the microphenomena with respect to phenomena on the macro level? Is it not so that in the macrophysical limiting case the quantum mechanical laws are reduced to the laws of classical physics, and therefore the picture of the world remains deterministic on the macrophysical level? The indeterminism of quantum physics should then not have anything to do with man's free will.

P: It is very important, in principle, that the quantum mechanical laws are in macrophysical cases reduced to the laws of classical physics. This is the requirement for the development of the new atomic theory that Bohr emphasized in the 1920s using the term *correspondence principle*, and it is one of the cornerstones of the Copenhagen interpretation. One should not, however, conclude from this that the picture of the world remains deterministic on the macrophysical level. In no way can one claim, for example, that spiritual life can be reduced to physics. Not everything can be reduced to atomic physics although in physics this has been possible to an unexpected extent.

The fact that one is forced to abandon determinism in atomic physics and to replace it with the idea of statistical causality should give reason to think of an analogous change in the other fields of empirical research. In no other field is there more motivation for the idea of an absolute causality or determinism than in physics. Thus, if determinism is not valid in physics, how can one think that absolute causality could be justified in the less exact fields? The validity of absolute causality is, in any case, a matter of belief (or should we say: a hypothesis), and experiences in physics are not encouraging in this respect.

Actually, the question really concerns the fact that our observations never uncover an object exactly as it is, but only give a picture of reality which is limited by the accuracy of our senses and by our shaping capacity. The unconscious processes which precede our 'becoming conscious of something' create a 'veil' which hides reality-itself from our knowledge. This is a fact which by no means concerns only atomic physics but empirical knowledge in general. From this point of view, determinism seems to be a Utopian idea. It is more natural to think of the world in the light of statistical causality than to believe in determinism.

This is how Bohr and Pauli judged the situation, as did Heisenberg. All of them thought that one has to abandon determinism in all fields of human knowledge, with the exception of pure logic. (Even there one has, in fact, met analogous problems; I refer here to Gödel's results concerning mathematical knowledge.)

S: I am beginning to understand that this is a question of a truly profound scientific revolution – a change of paradigm which refers to all empirical knowledge. With respect to the 'philosophy of one world' which you mentioned, I am, however, wondering about the difference between the organic and the inanimate nature. I am not worried about the new blow to the special status of human beings in nature which this seems to mean if the human 'soul' is associated with 'matter' in the same way as the 'souls' of the other organisms. I have got strong influences from the biological theory of evolution and don't think that human beings are so different from the other organisms. But what about inanimate nature? Do you think that even an elementary particle has a 'soul'? Do you end in some kind of panpsychism in your thinking?

Panpsychism

P: When I have illustrated the idea of statistical causality with the aid of the diffraction of radiation in narrow slits (see frontispiece to Part I), people often ask whether I think that a particle then has a 'will' so that it can freely choose its place on the film. Here we have a case where individual events are scattered around the main behavior, and in each individual case one of the possible places on the film 'is chosen': seemingly, the particle freely chooses the place where it leaves its visiting card on the film.

Pauli has indeed spoken of a *will* which acts in nature; the 'choices' in individual events are its expressions. Whether we should speak of the 'will' of an individual particle or of a 'will of nature' remains open here. In any case, there seems to be in nature an ability for choices, which is excluded from a deterministic picture of the world. One can very well describe the situation by saying that every individual particle has a 'freedom of choice' with respect to its final place on the film. In the case of inanimate nature, this choice seems to be 'blind', it is true, without any purpose or comparison between different possibilities, whereas the latter are characteristic of the choices made by living organisms.

However, freedom always has some restrictions. Different particles make their 'choices' independently of each other – Pauli has compared them with the 'windowless monads' of Leibniz which do not know anything of each other. Nevertheless, the total distribution of points on the film is quite regular and can be governed by a rational theory. The irrationality of reality which appears in those free choices in individual cases is restricted by a rational law concerning the total distribution.

S: But then this is a very mystical situation indeed. Is it really so that the different particles do not influence each other's behavior in any way, and yet the total result is governed by a law? There must be an explanation for this regularity of the total result!

P: At the very bottom, it is a mystery that there are laws of nature. According to Einstein, "the most incomprehensible thing about the world is that it is comprehensible." Now as laws of nature are found to be statistical, we meet another mystery: how can the whole be regular if individual events are 'free', without following any laws? It is a mystery which does not have any rational explanation – just as there is no explanation for the existence of the laws of nature.

S: But there must be some explanation of why points that are chosen quite randomly form a regular whole at the end!

P: There is no explanation. It is a mystery. We must be satisfied by just stating that the whole is regular although the choice is free from the point of view of each individual event.

Of course, physicists have passionately tried to find explanations. For example, the new approaches which I mentioned earlier are such attempts. I think that the enormous eagerness to find an explanation is caused by the fact that rationalists cannot allow any mysteries: They instinctively fear that we could land in the realm of religion.

S: In fact, we should speak of panpsychism which, in itself, is a very dangerous theme. From what you now said, I get the impression that the idea of the 'soul' of an elementary particle is not an entirely impossible idea for you. For me, however, it is still difficult to think that a particle 'chooses' and 'wills' something.

P: Such expressions have to be understood only as some kinds of metaphors which are used in order to describe situations where a better description is not possible. I remind you that we are now within the realm of metaphysics, and the only language we can use is the 'episteme'. Perhaps it will be possible to find a better description for these matters in the future – for example, if it is possible to find another basic principle on a par with causality. C.G. Jung tried to find such a principle in his conception of *synchronicity*, as did Leibniz when he spoke of the *preestablished harmony* (harmonia praestabilita) which combined the 'windowless monads' into an organized universe.

The main thing is that the decline of determinism opens a new degree of freedom for science; this has been striven for over the ages, but Western rationality has pushed it out of sight. It opens a perspective on the freedom of will in a metaphysical sense by eliminating the ontological obstacles which determinism in modern times has created.

S: But you have not answered the question whether you think that an elementary particle has a soul! Has it consciousness and free will?

P: Every answer to such a question can only be a metaphor. Anyhow, it is good to try to argue such matters.

In the most exact way, I can try to find an answer again in quantum mechanics because there we have a language for the description of microphysical events. When we use the state function for making predictions concerning a given system, we try to anticipate how this system will behave in a given situation. The state function must explicitly describe the situation in question. This is the only way of handling micro-systems, and it can best be described by saying that the system will in its behavior take into account the situation in which it finds itself. This was again a metaphysical expression but I cannot describe this situation better. In a certain sense, 'the system behaves in a purposeful way' in the situation in which it finds itself.

I would like to say that the microphysical way to describe a system's behavior contains in a primitive form a feature which is characteristic of organic nature: An organism tries to react to its neighborhood in a purposeful way – it tries to pay attention to essential features in the neighborhood. This is a feature which perhaps justifies speaking of an elementary consciousness.

'Choices' and 'will' concern the interpretation of the state function. As is known, the state function is a probability function, which makes it possible to present probabilistic predictions concerning the behavior of the system; in general it admits several different possibilities. For these one can calculate probabilities with the aid of the state function. If in an experiment we examine the behavior of one individual system, the irrational choice takes place at the end of the experiment. One of the possibilities 'actualizes' (materializes), and we cannot say which of them it will be. It is true that we can give probabilities for the different results, but in the case of individual events, probabilities do not mean anything: the choice 'made' by the system is 'completely free'. Only when we perform an experiment using a great number of similar systems in a similar situation – when we have sufficient statistics at our disposal – can we then verify or falsify the predictions which the state function gives us.

Here we meet the 'free choice' of the system in question. It concerns an individual experimental event. In this sense we can say that the system has 'consciousness' and also 'free will'.

S: This is a very strange story indeed. If one thinks of the matter in this sense, it looks as if the element which you wish to call 'psychic' would also appear in inanimate nature. In this sense I can see some idea in panpsychism. Then we do not need to worry at which phase of evolution 'consciousness' and 'free will' have come in. They are aspects in all kinds of 'matter'. Perhaps even the border between the 'living' and the 'inanimate' is not so absolute as one usually thinks. Evolution is just a gradual development towards a more and more complete con-sciousness and freedom, and it is not necessary to deliberate when and at which phase the step to the next higher level takes place: from the inanimate to the living and from the unconscious to the conscious.

P: I think that we can now leave the discussion of panpsychism and at last concentrate in the important question concerning creation.

S: Just now I have realized that these matters are closely associated with creativeness. One speaks of it so much today, and simultaneously people strongly criticize Western rationalism, as you also do when you emphasize the irrationality of reality.

Creation

P: Statistical causality describes phenomena in a way which can be called continuous creation. I will now assume that reality is by nature the way the genuinely probabilistic laws presuppose it to be. Every change then contains a choice between different possibilities. This is, in fact, a creation process because the result of every individual change contains something that cannot be predicted.

If we think of human creative activity from this point of view, it is natural to call the choices contained in changes acts of will which give direction to the changes. However, here we encounter reason as another factor because one cannot make choices in an arbitrary way, but instead deliberately – if one wishes to accomplish something permanently valuable. Creation is not only willing, but a collaboration of will and reason. (Here reason represents rationalism, an ability to see regularities in the world – laws of nature.)

This kind of characterization of creation still lacks something essential, of course: feeling. This is the power which produces activity; it gives rise to impulses for creative functioning and also performs the instinctive comparison between different possibilities. It is, however, such an 'irrational' factor that I am not able to say much about it – I am, after all, an exceptionally rational man although I have fought much against excessively one-sided rationalism!

Therefore we must be satisfied with this brief mention of feelings, which does not mean that I would judge them lightly. They are a conditio sine qua non – a necessary condition – for creative activity.

Thus, I would like to describe creation as a cooperation of reason and will: as deliberately performed choices. A conception of reality which includes a rational and an irrational basic aspect provides a natural foundation for the understanding of creativity. This is pertinently described by statistical causality. A statistical law always includes the irrational freedom which is a precondition for real changes, but freedom, on the other hand, is restricted by the statistical law. If freedom is used without considering the regularities (laws) which always have an influence in the background, the change remains an exceptional event without any significance.

S: The matter-of-fact laws of nature here are, indeed, applied to very poetical matters because I suppose that by creativity you also mean creative processes in the arts. Certain aspects of creation, in fact, are described very nicely in your way; I find it actually a little surprising that statistical causality can be used even for this kind of purpose.

The collaboration between reason and the creative will which you have emphasized here sounds rather progressive today when many people speak so eagerly of putting more emphasis on creativity in schools, while the present school system too one-sidedly lays stress on intellectual development. However, don't people now speak a little too much of creativity, after all?

P: I think that the one-sided rationality of Western culture is criticized with good reason. The firm opposition to the idea of the irrationality of reality has been illuminating for me in this respect. The idea that free choices really take place in individual events, without any motives or causes, seems to be impossible for most physicists and philosophers. These are, however, basic ideas in Pauli's philosophy, as far as I understand it. Rationality seems to be so deeply rooted in Western people that one must find an explanation for everything 'using reason correctly'.

However, speaking of creativity is often very superficial in that one does not stress enough its being subjected to discipline. Freedom can be misused, and then it does not produce anything. It is meaningful only from the point of view of a certain rationality, a certain regularity. There must be a basic idea which indeed means a regularity (law), and freedom has importance only against the background of this regularity; in the case of statistical causality, this is the law concerning the behavior of a great many individual events as a whole. One can also learn creativity by solving mathematical problems, and then it is associated with a very strict discipline.

Anyhow, it is important to realize that creativity presupposes indeterminism, freedom from mechanistic laws. It is associated with an essential irrationality which simultaneously means freedom. For me, the comprehension of this irrationality and of the free choices has been an extremely strong positive experience which has had an essential influence on my thought. When I clearly realized their importance for the first time, I was in a certain crisis in my life – I had also had a slight heart infarct – and I experienced it as an extremely soothing feeling of *God's presence in everything that takes place.*

I beg your pardon, but now I am going to neglect your wish that we should not speak of religion today because I cannot make these things clear without referring to some religious matters. To me, God has spoken exactly in this way, and it has been the kind of talk which truly has had an influence on me, since a talk referring to reason creates the strongest resonances in me. When I, *on the basis of rational deliberation*, became convinced of the fact that there are matters which must be called real and which, however, cannot be explained by reason, this was a turning point in my thought.

S: I hope, however, that we remain on a basis where reason has not been lost.

P: When I have tried to understand the deepest motives of the attitude of my colleagues, I have repeatedly clashed with *belief.* Most of them repress religion so strongly that this repression forces them to another decision of faith resulting in scientism: to the confidence that all real problems can be solved by scientific methods. What I just said means that I then became free of scientism and instead met God, who is present and has influence here and now. And step by step it became clear to me that this God, whom I met in the heart of nature, is the same as the God of Christianity. I began to understand the Bible in a new way. It became living for me through the witness of nature.

In several connections I have described some details of what I have found here. I would now merely like to add some words about creation – about the creation of the world. The continuous creation which we have discussed here, in fact, belongs to the creation of the world. When man creates something new, he can feel that he is like an instrument in the hands of some invisible power. There is a living, creative force effective in the world, and I feel that I am its coworker when I try to formulate a new idea or try to get a new plant to thrive in a place in the garden where it seems to fit into the general plan. This is the 'teleological factor' which I described in Chap. 3 on "Scientism".

As a physicist, I comprehend matter as a very abstract thing. The fundamental constituents of matter appear to be quite different from what we usually mean by matter. The quarks and interaction quanta are not hard or soft, and they don't have any color. Characteristic to them are certain conservative – invariant – properties such as electric charge and spin. Invariant properties are, on the other hand, expressions of symmetries which are characteristic of the structure of the 'particle' in question and its interactions with other 'particles'. In fact, the fundamental properties of matter are those symmetries which appear in experiments as the conservation of certain quantities.

In general, the 'particles' are not at all permanent but can be transformed into each other and create quite new 'particles'. Only those invariant properties of the 'particles' are permanent, and what is permanent can depend on the circumstances.

The nature of elementary particle processes means a continuous creation of something new: Each individual process contains the irrationality which we are discussing. Natural laws do not determine unambiguously what will occur: There are always possibilities for choices. There is always space for the creative will. Therefore, attempts which rationalists make to describe the birth and the evolution of the world on the basis of mere physical laws are untenable as a matter of principle.

S: Very authoritative physicists have, however, developed purely physical theories about the evolution of the world, and I have understood that physicists in general take them very seriously. These theories do not contain any irrational factor or any creative will about which you speak. It seems that you are pretty much alone with your remarks, aren't you? The most important of your remarks is perhaps the idea that evolution takes place in the form of a stepwise, stochastic process. Therefore, you can point out that there is an irrational factor included in each step, and to this factor you give, in addition, a teleological meaning. What do the other physicists say about such ideas?

P: Reactions have been supercilious. So far I have not heard any relevant remarks, only details concerning the Grand Unification Theory which should help me to understand the hard facts. My colleagues seem to be so imbued with scientism that they cannot understand at all that I am criticizing the belief which is the basis of their approach to these questions and of the interpretation of the 'facts' – that I cannot find the purely rational argumentation in these questions reliable.

I must again and again point out that the essence can be found in the individual events. In a theory employing continuous functions, this basic aspect is eclipsed, especially if one speaks of the 'state function of the universe'. This concept is not possible according to the original 'Copenhagen philosophy'. The 'state function of the universe' presupposes explicitly the idea of an objective reality which is independent of observations. Such a concept is not allowed in a purely empirical science because an empirical science can only use concepts which are defined on the basis of observations. *An objective reality which is independent of observations is a metaphysical concept.* Some kind of vision of the nature of such a reality is then necessary, and this is an expression of some belief – for instance, a materialistic or rationalistic belief.

The discrepancy between Bohr and Einstein explicitly concerned the idea of an independent reality. According to Bohr, it is not possible to speak of a reality independent of observations, and this critical attitude is certainly justified. Einstein, however, believed that there is an objective reality which is completely rational by nature, and this belief guided him to certain views concerning the microreality which were not compatible with the Copenhagen interpretation of quantum mechanics. Because of this, he claimed that quantum mechanics does not describe reality completely. Later, the EPR experiments clearly showed that Einstein's rationalistic belief guided him to associate with reality properties which do not correspond to experimental results. Therefore, Bohr's cautious attitude towards the idea of an objective reality, independent of observations, has been shown to be well justified.

Since the 1960s, rationalists have made more and more attempts to describe the independent reality (the universe) on the basis of the Einsteinian belief that reality can be described purely mathematically. Obviously, people presuppose that science must be based on scientism, i.e., that all problems concerning reality must be solvable by rational methods. I have tried to point out that this belief is not reliable and that atomic research does not seem to support it. So far, no relevant discussion has resulted. This I find to be very clear evidence that certain beliefs are presupposed in science. This means that *science has become like a religion*, and therefore it is understandable that people generally think that religion and science are incompatible.

S: To me, religion has remained rather unfamiliar, and I must say that I find it strange that you claim to be able to combine the honest scientific world view with Christianity. Even if you speak of a continuous creation in the form of a guiding, teleological factor in phenomena, this should not mean that you believe in

Genesis in the Bible and more generally in the many fanciful stories there. Since we have now come hopelessly into the realm of religion, I should be allowed to ask you: In what do you believe, in the end?

P: You are allowed to ask but unfortunately you won't get such detailed answers as you probably expect. I mean, for me religious belief does not mean belief in certain facts in the sense that one speaks of facts in science. I find it meaningless to require that every Christian should believe that there is a common father for all people, in the form of man, who with the power of his word created the world from nothing in six days, i.e., 6 x 24 hours. I am searching the Bible for things other than a scientific picture of the creation and the evolution of the world. I believe in God, who has created the world and is the basic origin of existence. He is also a ubiquitous, living God who continuously influences what takes place. However, this origin of existence is transcendent, not understandable by human reason. Whatever we imagine about God is certainly a wrong vision.

For me, God is the same as reality itself, its most profound essence. I do not mean here the universe which science describes. Reality is deeper; it is 'beyond' the universe that can be investigated by human reason. The metaphor that man has been created in the image of God means to me that we can, by using human understanding, at least to some extent, comprehend this reality where we are living. I think, as Einstein did, that by studying nature we can learn to know something of 'the Old', whose works we can only admire. However, man is not an intellectual being only. Not even reality is as a pure-blooded rationalist imagines. Reality is not a machine. It can better be compared with an organism which lives and reacts in some 'purposeful' way. But in general we do not know this purpose. There is irrationality inherent in reality, not reachable by reason. It possesses creativity which continuously creates something that did not exist before. Even these basic features of reality reveal something about God.

S: But those are things of a very general nature and not impossible to accept. This does not yet mean that you believe in a triune God (which in itself is a contradictory concept), in a creator of the world who sacrificed his own son in order to redeem us from hell which he himself created and to which he, in the end, sentences most of the people he has created. For me it is difficult to imagine that this kind of religion could be vital continuously. We have, indeed, learned to be critical in science and are able to some extent to judge which statements are reliable and which are not.

P: Now I must request that we do not speak so much of religion. In this single discussion we cannot clear up all of the Christian teachings and how they should be understood. For example, the doctrine of the Trinity contains a really fine idea – from the point of view of the 'correct use of reason' if we don't limit the use of reason to the strictly rational knowledge which I have characterized by the Platonian term *dianoia*. If we accept the more general use of reason which is needed in metaphysics, i.e., the knowledge in the sense of *episteme*, then the idea of Trinity contains a very profound wisdom which opens important perspectives, but it is done by using the language of metaphysics. The discussion of the doctrine of Trinity would, however, require another meeting. It also opens a view to creation which we are now treating, but the perspective is quite different from this discussion, and it is better to leave it for another time.

Thus, let us further continue the discussion of creation from the same point of view as we have done today and 'on a basis where reason has not been lost', as you stated.

S: One question still remains unclear to me. Is it not possible to investigate experimentally whether the distributions of individual results in quantum mechanical experiments correspond to the normal distributions in the theory of probability? If they do, then the attitude which you call the 'religion of chance' would be experimentally justified. If there is a teleological factor influencing certain processes, it should be possible to show it by demonstrating that the distribution of individual events deviates from the normal distribution which presupposes purely accidental events!

P: It seems that people have considered this a crushing argument against teleology. At least I have heard some physicists state that in this way the 'finger of God', which according to some people steers the motion of particles, can be shown to be nonexistent.

I think physicists in general believe so because they have an unshaken faith in 'pure chance'. In the case of so-called inanimate nature, I truly don't claim anything else than that distributions there in general correspond to the normal probability distributions of accidental events. The issue is changed in the more and more conscious processes in living organisms. In these processes, the investigation of distributions can be impracticable, especially because an organism reacts in a more or less abnormal way when it realizes that it is being examined.

The investigation of such problems should interest biologists, and I think that the expressions of teleology in the behavior of living organisms deserve more attention they seem to receive. An outsider gets the impression that people avoid the investigation of teleology because otherwise a scientist can easily become a target of supercilious criticism similar to that which I have encountered among physicists!

The 'religion of chance' leads, as is well known, to the idea that the universe as a whole develops toward an equilibrium which is called heat death: a state where all thermal differences have disappeared and no changes are possible any longer. Teleology, of course, takes away the basis of such a conclusion. If we believe in teleology, there is will effective in all changes, and this is able to increase order while the second law of thermodynamics (the entropy law) states that the processes always proceed towards disorder. It is remarkable that the evolution of life, in particular, has proceeded in a direction where the degree of order has been increasing and not decreasing!

Will is indeed a factor characteristic of life – a creative factor. It gives a different direction to the cosmological considerations than a perspective which is based on the processes of inanimate nature.

The way a rationalist describes cosmology is based on the metaphysical belief in a reality which is independent of observations and on the belief in the complete rationality of this reality. The empirical justification of these beliefs is based on the good predictions which have been obtained on this basis. In atomic physics these ideas have met severe difficulties, however. If one continues developing theories on this basis, this can only be based on a belief, not on an empirical justification. But in the realm of belief, 'truths' must be evaluated using grounds other than those used in science.

The belief in creation is an inseparable part of the Christian faith, and it is dangerous to imagine that it would be possible to describe creation as a purely rational process. The world is not a machine but, rather, an organism, and besides rationality, phenomena contain the 'irrational' factor of will. This is characteristic of life. *Rationalism leads to a cosmology where the idea of creative will is missing and, therefore, the world is without life.* In this way, an imagined universe results; it is not the world in which we humans are living as conscious beings searching for truth. From that imagined world one does not find any God. In our real world an honest seeker after truth cannot avoid encountering Him.

10. Numinosum

Flashes

There is, for example, the case of the theologian which I described in "Archetypes of the Collective Unconscious." He had a certain dream which was frequently repeated. He dreamt that he was standing on a slope from which he had a beautiful view of a low valley covered with dense woods. In the dream he knew that in the middle of the woods there was a lake, and he also knew that hitherto something had always prevented him from going there. But this time he wanted to carry out his plan. As he approached the lake, the atmosphere grew uncanny, and suddenly a light gust of wind passed over the surface of the water, which rippled darkly. He awoke with a cry of terror.

At first this dream seems incomprehensible. But as a theologian the dreamer should have remembered the "pool" whose waters were stirred by a sudden wind, and in which the sick were bathed – the pool of Bethesda. An angel descended and touched the water, which thereby acquired curative powers The light wind is the pneuma which bloweth where it listeth. And that terrified the dreamer. An unseen presence is suggested, a *numen* that lives its own life and in whose presence man shudders. (*Jung* 1961, p. 163.)

The term *numinosum* was introduced by the German theologian and religion psychologist Rudolf Otto to describe the experience of the deity (*Otto* 1917). In Latin, *numen* = deity, majesty, the divine will.

Maybe I can best amplify my description by a passage of Scriptures which once, in a moment of despair, opened to me from the Bible, illuminating, like a flash of lightning, the dark night:
> If ye had faith as a grain of mustard seed, ye might say unto this sycamine tree, Be thou plucked up by the root, and be thou planted in the sea; and it should obey you. (Luke 17.6.)

(*Laurikainen* 1978, p. 193.)

In the beginning was the Word, and the Word was with God, and the Word was God. (John 1.1.)

The Greek equivalent of Word is *logos*. Heraclitus, who was a philosopher of change, used the term logos to mean the invariant essence of change – equivalent to what in physics is called the law governing change – or a superhuman intelligence which is the origin of the rationality of the world. For Stoics, Logos was a god who created the rationality of the world.

In Platonic philosophy, the correspondent to logos is *that which truly is*, the world of ideas which is imperfectly materialized in the visible world. For the neo-Platonians, logos meant the rational basic form of the world; it was a mediator which gave a picture of the One, the transcendental basis of existence, in a form comprehensible to human reason. Thus, Logos is an intercessor between the visible world and the transcendental One which is beyond the reach of our reason: the Son of God which gives us an idea of the incomprehensible God.

Holy Trinity

Pavel Florensky is a representative of the orthodox theology of this century and, simultaneuously, of the philosophy of religion; to orthodox Christians these two mean the same. The mystery of the Holy Trinity has a central place in Florensky's theology. (See *Silberer* 1984.) For me, it has been an encouragement to penetrate into the difficult problems concerning Trinity. It is perhaps strange that number mysticism and Jungian psychology have helped me to understand them better.

Florensky was a mathematician and physicist by his academic education, and during Stalin's time he attained a respected position. However, he also had an orthodox ecclesiastical schooling, and alongside his profession, he worked as a clergyman. Before long this provoked criticism. Florensky ran into difficulties: Finally, he was sent to a concentration camp where he died some ten years later.

The theological views of Florensky have been presented mostly in letters – partly written to an imagined addressee. They are attracting more and more interest now, both in orthodox circles and elsewhere.

At the Council of Nicaea in 325, the creed based on the idea of the trinity was adopted. This included pushing aside Gnosticism and the Arian heresy. According to Arius, the Son or the Word was not eternal but created although prior to the other created beings. In the Nicene Creed it is stated:

We believe ... in one Lord Jesus Christ, Son of God, the only-begotten, born of the Father, that is, of the substance of the Father, God of God, light of light, true God of true God, born, not made, of one substance with the Father, through whom all things were made.

According to this creed, the three divine persons or hypostases are *of the same substance*. This trinity is the fundamental mystery of Christianity. The Son is the Word of God, of the same substance as the Father, who became a man in order to glorify the Father in a way which is understandable to human beings.

The Trinity is a mystery. According to Florensky, it is not a thing comprehensible by reason alone: In order to understand it, reason must search for its own roots. *Faith is primary*; it gives direction to understanding. Faith develops gradually from a blind belief into a conscious belief which is in harmony with understanding; then God is not only a subject of faith but also a subject of knowledge. Then there is no longer any strict boundary between faith and knowledge but each of them presupposes the other. This most complete form of faith Florensky describes by the principle: "*Intelligo ut credam!*" – I understand in order to believe.

Thus, it is important that we also understand the trinity although this understanding is not a thing of pure reason. The starting point for the understanding is the divine substance (*ousia* in Greek) which is the absolute truth and the absolute love: *the substance of God is truth which appears as love.* The spiritual life which goes out from the Father and which has Him as its center is by nature the understanding of the truth.

Understanding presupposes an object which is the subject of the truth. Therefore the divine I presupposes a divine Thou, the object of understanding. God's substance presupposes that I gives birth to Thou.

From the point of view of the divine life – "seeing from inside" the trinity – understanding means giving birth; "seeing from outside" it means creation. So the Father has, according to His substance, before the beginning of time given birth to the Son, who is eternal like the Father. The Father is the basis of existence, but the substance of the Father presupposes the Son, who is the object of understanding.

Thus, because I and Thou exist, *they must have some relation to each other.* This relation is the third person, 'He' or the Holy Ghost, and *its substance is love.*

This involves the internal dynamics of the Godhead. Its substances are truth and love. Truth means understanding which gives birth to a Thou for the I, and love binds them into a unity.

The trinitarian theology describes in detail this inner dynamics of the Godhead and its relation to creation. The idea of the Creation reflects the birth of the Son from the Father. This is the heart in the dynamics of the trinity, not comprehensible through pure reason.

The trinitarian theology of Florensky emphasizes very strongly the same substance (consubstantiality) of all the divine Persons; it is *homoousian philosophy.* Thus, love belongs to the substance of the Father as well, and when the Father gives birth to the Son, it is not only an expression of the truth-aspect of His substance but, simultaneously, an expression of love.

If one tries to characterize the different divine Persons with the aid of the basic attributes of God, then characteristic of the Father is *existence:* He is the origin of existence. Characteristic of the Son is, in the first place, *love* because the Son is the person who by becoming man and by His death gives witness of God's love toward the creation. The characteristic attribute of the Holy Ghost is, according to Florensky, *beauty;* the Third Person makes the inner dynamics of the Godhead a beautiful whole.

The role of the Holy Ghost in the divine dynamics is further clarified by Florensky's remark that the Third Person can better be characterized by the pronoun 'We' than 'He'. The Holy Ghost is the *hypostasy which forms wholes.* In the creation as well it appears as a factor which creates the unity of the congregation, the communion.

According to the Tomists, the fundamental mystery of the Godhead, where the Father gives birth to the Son, is more strongly related to 'truth' than to 'love'. Florensky points out that 'existence', 'truth', and 'love' belong to all Persons of the Godhead (this is an expression of the *homoousia*). Then 'love' becomes more strongly emphasized in the dynamics of the Godhead. The Tomistian view of the trinity is more static, while Florensky emphasizes strongly the dynamics of the Godhead which in its relation to the created world appears in the *act of creation*.

Number Mysticism

The idea of trinity is very close to the Pythagorean and the Oriental number mysticism. In number mysticism, integers have qualitative properties in addition to their arithmetic relations, which can be described in a strictly rational language; the qualitative properties can be understood only in a metaphysical sense, using the language that Plato calls 'episteme'.

The nature of the number *one* itself presupposes that there is something more than this being one. This is the idea of *two*. But if there are the beings 'one' and 'two', they must be in some relation to each other, and this relation is the idea of *three*. If 'three' is understood in this way, it combines 'one' and 'two' into a whole – and this is similar to the idea of trinity.

The fundamental mystery is how *one* 'gives birth' to *two*. This mystery is inherently associated with the concept of *consciousness* because 'consciousness' presupposes that there is something that is its object – something of which one is conscious. That we 'become conscious of something' presupposes that the object is separated from the subject. 'I' presupposes in some form 'Thou'. In the dynamics of the trinity, this idea is the same as the idea that the substance of the Father presupposes the existence of the Son.

This leads quite logically to *three* because if there are two beings, it is natural to ask what their interrelation is, and this relation is already a third logical object.

How one comes to number *four* can cause very mystical speculations which lead outside of trinity – *from trinity to quaternity*.

One should, of course, be careful in not mixing metaphysical concepts with a strictly rational theory. Such a metaphysical concept is the idea of a *reality independent of observations*. If we wish to remain in the realm of empirical science, the existence of such a reality remains a question of belief – as is the applicability of number mysticism.

Irrationality

Characteristic of quantum mechanics is a certain *discreteness*: Quantum states are clearly distinguished from each other, and 'quantum jumps' from one quantum state to another are some kind of 'process-atoms' which cannot be subdivided in any way. On the other hand, the formalism of the present quantum mechanics uses continuous functions and, correspondingly, space and time are assumed to be continuous. Quantum theory is characterized by a tension between discreteness and continuous theory. In fact, the probabilistic interpretation of the theory is a consequence of this tension, and the irrationality of reality also follows from it.

Florensky points out that science has now met the idea of discreteness, whereas previously it had been bound to the idea of continuity. He considers this a very profound change which is closely related to the mysticism of the trinitarian theology; therefore, the idea of discreteness is in contradiction to contemporary science which is based on the idea of a complete rationalism.

According to Florensky, the nature of the Third Person of the Godhead has not become sufficiently clear in the present theology. At the center of interest has been the Second Person: Son of God or the Word – Logos. It has led to an overemphasis on rationality and also to a purely rational science. The appearance of discreteness in science Florensky finds to be a very profound change, and he anticipates that, as a parallel phenomenon, the conception of the nature and the functioning of the Holy Ghost will become clearer in the future theology.

The strong opposition to the irrationality of reality is in harmony with Florensky's views. It is a question of truly profound change in the basic direction of science which, in fact, concerns the *basic faith supporting Western culture.*

In any case, it is important that Christianity explicitly contains an irrational element. Love cannot be dominated by reason. Florensky's remark that the present science is based on a belief which emphasizes primarily the Second Person of the Godhead – the Word, or Logos – is very interesting. Perhaps we need a new science which more clearly takes into account the Third Person which binds the Trinity to a whole, i.e., a science which is associated with a faith which is 'foolishness unto the Greeks'.

For the preaching of the cross is to them that perish foolishness; but unto us which are saved it is the power of God.

For it is written, I will destroy the wisdom of the wise, and will bring to nothing the understanding of the prudent.

Where is the wise? where is the scribe? where is the disputer of this world? hath not God made foolish the wisdom of this world?

For after that in the wisdom of God the world by wisdom knew not God, it pleased God by the foolishness of preaching to save them that believe.

For the Jews require a sign, and the Greeks seek after wisdom:

But we preach Christ crucified, unto the Jews a stumblingblock, and unto the Greeks foolishness;

But unto them which are called, both Jews and Greeks, Christ the power of God, and the wisdom of God.

Because the foolishness of God is wiser than men; and the weakness of God is stronger than men.

(1 Cor. 1.18–25.)

11. The World of Spirit

Transcendent Reality

Theologian: This is certainly a very quiet corner of the restaurant. Do you think that our waitress will mind if our orders are light and we just concentrate on philosophy?

Physicist: I have explained the situation to her, so don't worry. It was very nice that you found time for this discussion because I really need your expertise. When studying atomic physics I have come upon so many different matters that I need criticism from experts in different fields in order not to go astray.

T: Thank you very much for the manuscripts and reprints you sent to me. I have read them with great interest, indeed, but I must say that I do not know enough about the matters which you discuss in them. I need elucidation on several details. I hope that stupid questions are allowed, because my knowledge of physics is extremely minimal.

I understood that we should continue your considerations about the doctrine of the Trinity and especially discuss the relation between God and the Creation and the role of the Holy Ghost in this respect.

P: I would like some additional illumination on these questions. Theology is a foreign field to me, and I sorely need help and criticism from those who know the field. I am a complete layman in these questions. However, Florensky's thought on the one hand and, on the other, writings by Polkinghorne (see, e.g., *Polkinghorne* 1988, 1994) and articles concerning lectures at the European conferences on science and religion (see, e.g., *ECST* 1990, 1994) contain views which are quite close to ideas which I know very well from the philosophical problems of quantum mechanics. I would like to analyze these connecting points in more detail. Particularly, the role of the Holy Ghost in the doctrine of the Trinity and in God's relation to the Creation interests me much now.

T: You mentioned that proceeding from 'three' to 'four' leads outside the idea of the Trinity: from Trinity to Quaternity. Does it mean mixing Christian belief with some ideas borrowed from other religions, some kind of trend toward a syncretistic religion? On the basis of the knowledge I have about the syncretistic religions, I must be critical with respect to such trends.

P: In my case, you don't need to be worried about syncretism. I understand well that religion has deep roots in the human personality, and the mixing of dogmas from different religions can only be some kind of intellectual game. In such a way one does not find any real answer to questions that a human being searches for in religion.

For a theologian, it probably sounds very suspicious that I have compared the trinitarian theology with number mysticism. In Christianity, the doctrine of the Trinity is considered a mystery which concerns a great and properly Christian secret. I understand it differently. I cannot take even Christianity as a revelation religion where the Bible is a collection of holy truths which must be accepted without criticism. In science I have learned to criticize everything. I am searching for verification for everything from the world of experience in which I live, and even the truths that Christianity offers to a seeker of truth I can only adopt on the basis of such deliberation. The fact that the basic ideas of the Trinity can also be found in number mysticism strengthens my belief in this mystery because this is an indication of its more general importance.

The number theorist Leopold Kronecker has said, "Integers has God created, everything else is made by men." This aphorism strongly emphasizes the basic role of integers in shaping a picture of the world. Mathematical natural sciences as a whole are, in the end, based on integers, and, on the other hand, integers reflect in a profound way the functioning of our psyche.

Number mysticism, in a manner, is an expression of the way the human psyche gives shape to reality. Basic ideas of number mysticism are found in different cultures. It has also given birth to the completely rational arithmetic, and this is the only use of integers which Western science accepts.

It is wrong, however, if number mysticism – the discussion of the qualitative properties of integers – is rejected only because it does not fulfill the requirements of exactness used in rational science. In its own way, number mysticism opens a new perspective into the world of numbers, simultaneously illuminating the functioning of our psyche and the basic shapes of reality which are materialized in the structure of the material outer world. Integers open one of the most exact possibilities for gaining sight of the mysterious world of archetypes which order the unconscious functioning of the psyche.

T: You have a great interest in the depth psychology of C.G. Jung where those archetypes play an important role. I found Jung very difficult to understand, and I have heard critical remarks from our psychologists concerning the inaccuracy of Jung. For example, the definition of archetype seems to remain very obscure and even self-contradictory. Do you think that it is a scientific concept? According to Jung's description, archetypes seem to be something hiding so deep in our unconscious psyche that it is impossible to define them. Archetypical images or ideas are more easily understandable, but then Jung again begins to speak of archetypes as something that do not belong only to the psychic life but can be related to the material world as well. I think he has used the term *psychoid* in this connection. Is it not a little too unclear in order to be scientific?

P: I regard the concept of archetype as extremely important. It is true that Jung has given different definitions of archetypes in the course of time. In any case, archetypes are the most important connecting link between Jung's and Pauli's thought – for example, in their joint book *Naturerklärung und Psyche* (1952). Over the years, there was a clear development in the idea of archetype in Jung's thought, and it is possible that discussions with Pauli had a certain influence, in that this concept acquired a more and more abstract meaning. After 1950, Jung made a very clear distinction between archetypes and archetypal images. The archetypes themselves he describes as very abstract patterns of thought; only dim, intuitive ideas of them are possible on the basis of the archetypal images or ideas which they produce.

I understand archetypes as shaping abilities which are characteristic of the unconscious functioning of the psyche. They are modes of action of the psyche produced by our genetic or cultural inheritance. Archetypes shape the picture of the world in which we live and from which our senses convey to us different kinds of impressions.

A good example of the archetypal ideas which the archetypes produce are natural numbers or integers. With the aid of the integers, the shaping and ordering of our experiences becomes exact. Another example is mathematical group theory. It also contains very general concepts which can be applied in most different fields helping us to shape a picture of the world. Important applications of group theory are symmetries which can be found in most different connections both in nature and among the 'artefacts' produced by human beings. Group theory also has important applications in mathematics and mathematical physics. For example, the theory of elementary particles and their interactions can in essential respects be reduced to abstract symmetries.

T: Is it possible to understand Jung's idea of psychoid with the aid of archetypes in the sense that the most profound actions of our psyche are expressions of some 'patterns' which are common to both the material processes in our brain and the functioning of the human psyche (shaping processes, etc.)?

P: Oh yes, this was a telling characterization. Pauli considered reality to be psychical as well as material. We experience reality in the form of these *complementary phenomena*, and thus it can be called psychical as well as physical. This idea is characteristic of Pauli's thought – but also very difficult to comprehend. According to this view, physics and psychology are complementary sciences, or instead of physics one could speak of the natural sciences in general. Only together can the natural sciences and psychology give a reliable picture of reality.

I would like to call reality spiritual. According to an old Christian view which appears in some other traditions as well, there are three basic elements in a human being: *body, soul, and spirit.* The most fundamental element is spirit, which is immortal and the basis for communion with God.

There is great confusion in the use of the words 'mental' ('psychic') and 'spiritual'. The usage would become more consistent if we associate 'material' and 'psychical' (or 'mental') with the world of human experiences, using them for the description of phenomenal expressions of reality, while 'spiritual' would be reserved for a very abstract use: to describe reality itself which we cannot reach in our experiences.

T: Excellent! This is terminology which I would warmly recommend. The basic nature of reality would then not be describable using terms which are directly related to our experiences. 'Spiritual' means a very abstract reality which can appear to us either as 'material' or as 'psychical', depending on the object of our interest. 'Material' and 'psychical' are like projections of the 'spiritual' reality. These projections correspond to the two basic aspects of the human experiences – the 'outer experiences' and the 'inner experiences'.

P: We can say that there is a material and a psychical dimension in reality while spiritual is some kind of union of these dimensions. The psychophysical problem concerns the nature of the relation between the two basic dimensions of this union. The present science has a very dim view of this relation – or more precisely, no view at all. The psycho-physical parallelism tries to describe the material and the psychical dimension of reality as 'parallel' to each other. The material and the psychical are thus associated with each other like the two sides of a coin. This hypothesis of 'parallelism' is very obscure and weakly justified, however. (Descartes and Leibniz wrote that God has created the world in this way, but in modern science this explanation is not acceptable.) Only very poorly justified attempts at an explanation of the psycho-physical relation exist so far in science.

It is a fact, however, that we have these two aspects in our experiences, and they are in some way inseparably associated with each other. Matter and psyche form a very abstract union, and we decided to call it spiritual.

T: We could imagine that the world of our experiences is a two-dimensional plane where the material dimension and the psychical dimension span a coordinate system.

P: Exactly. In this way we arrive at a geometrical visualization of reality. However, we must remember that statistical causality implies the irrationality of reality. Thus, the world of experiences is open: It is as if it get 'influences from the outside', i.e., influences which cannot be explained by rational theories. In order to visualize this situation, we must think that *our plane is embedded in a three-*

dimensional space, where the third dimension, perpendicular to the plane of experiences, describes the 'irrational influences' – the irrationality of reality.

This three-dimensional space represents *reality* in this visualization. The plane of experiences is just an intersection of this space. Rational science operates in this plane: In principle, it tries to exclude all irrationality. (A critical analysis shows, however, that the absolute exclusion of irrationality is impossible.) Everything outside of this plane can be called *transcendent* from the point of view of the rational science. The third dimension cannot be reached by rational methods.

The term 'spiritual' now refers to the three-dimensional space. The 'irrational influences' can be explained, in principle, as 'influences' originating in the third dimension.

All in all, the world of experiences which science investigates is embedded here in spiritual reality. Thus, it is possible to imagine that this three-dimensional reality continually influences our experiences which then always have both rational and irrational influences, the latter having a 'supernatural' origin. An 'invisible reality' is present in the world although it cannot be revealed by rational, scientific means.

Man is also a citizen of the spiritual reality. Spirit is the nature of his *real self,* which we experience as material and psychical personality. We can think that death means detachment from the world of experiences, from the world of space and time. What this means in detail belongs to the secrets of spiritual reality. It means something that our rational thought cannot reach.

T: As a theologian, I do not see anything wrong in this kind of visualization of these matters if it can help some scientifically educated people to understand them. In fact, it begins to attract me.

P: I would still like to add something very important. For me this three-dimensional, spiritual reality is an *image of God*. That which truly is, is an expression of its origin – God. I must especially point out that the third dimension makes reality transcendent: It cannot be identified with the universe that science investigates. This picture does not mean any kind of pantheism. *God is transcendent,* not describable rationally.

However, simultaneously *God is immanent,* ubiquitous in the world of our experiences. The plane of experiences is embedded in spiritual reality. God's will – which is irrational by nature – influences everything everywhere. In this visualization, no contradiction can be seen in God's being simultaneously transcendent and immanent. We can also understand St. Paul's statement on Areopagus in Athens: "For in him we live, and move, and have our being" (Acts 17:28).

Of course, God is also the origin of all invariances, the rationality in the world. *God is the transcendent basis of both being and becoming.*

Perhaps a difficulty for theologians is the fact that all these considerations are based on physics. Since the Reformation, the general revelation has been mostly in the background, at least in the Lutheran theology. One is not accustomed to thinking that man can also learn to know God by studying nature. One should get rid of the fear of pantheism and pay more attention to the general revelation, because this can remove obstacles which make the Christian faith difficult for educated people to understand.

The question of whether God is personal must be solved on a quite different basis. I shall come back to this question in another connection. To me, personally, the analysis of the nature of observations has elucidated the fact that *God is a personal God who alone can open the door to truth.*

Let us, therefore, forget the fear of pantheism today and continue our search on the basis of physics. We have agreed on terminology which completely corresponds to my view concerning the words 'body' ('matter'), 'soul', and 'spirit'. 'Material' and 'psychical' are attributes referring to the world of our experiences while 'spiritual' refers to the transcendent reality which can only be an object of belief, not an object of rational science.

I would like to elucidate this further by quoting the first epistle of Paul the Apostle to the Corinthians, chapter 15:

But some *man* will say, How are the dead raised up? and with what body do they come?

Thou fool, that which thou sowest is not quickened, except it die:

And that which thou sowest, thou sowest not that body that shall be, but bare grain, it may chance of wheat, or of some other *grain*:

But God giveth it a body as it has pleased him, and to every seed his own body.

All flesh *is* not the same flesh: but *there is* one *kind of* flesh of men, another flesh of beasts, another of fishes, *and* another of birds.

There are also celestial bodies, and bodies terrestrial: but the glory of the celestial *is* one, and the *glory* of the terrestrial *is* another.

There is one glory of the sun, and another glory of the moon, and another glory of the stars: for *one* star differeth from *another* star in glory.

So also *is* the resurrection of the dead. It is sown in corruption; it is raised in incorruption:

It is sown in dishonour; it is raised in glory: it is sown in weakness; it is raised in power:

It is sown a natural body; it is raised a spiritual body. There is a natural body, and there is a spiritual body.

And so it is written, The first man Adam was made a living soul; the last Adam *was made* a quickening spirit.

Howbeit that *was* not first which is spiritual, but that which is natural; and afterward that which is spiritual.

The first man *is* of the earth, earthy: the second man *is* the Lord from heaven.

As *is* the earthy, such *are* they also that are earthy: and as *is* the heavenly, such *are* they also that are heavenly.

And as we have borne the image of the earthy, we shall also bear the image of the heavenly.

Now this I say, brethren, that flesh and blood cannot inherit the kingdom of God; neither doth corruption inherit incorruption. (1 Cor. 15: 35–50.)

The Trinity

P: Could we now, after this conceptual clarification, proceed to discuss the Trinity? I would like to learn what you think of certain views which interest me. But I must repeat that my views are born out of the discussion of the philosophy of physics; I am in unknown territory here and would not like to give rise to unnecessary confusion.

T: Here we are in a field where each one of us meets unfamiliar questions. I promise to listen but not to take a stand in regard to questions which I do not understand.

P: Once more: all my statements are a physicist's ideas, not theology. When reading Pauli's and Jung's articles and letters, I realized that in physics one can meet patterns of thought and problems similar to those in religion. Of course, one must be open to metaphysical questions – ready to treat the 'irrational territory'.

In 1948, Pauli wrote an essay "Moderne Beispiele zur 'Hintergrundsphysik'" [Modern examples of 'background physics']. It is now published as an appendix to the Pauli-Jung correspondence. It is an extremely interesting attempt to illuminate the unconscious background from which the concepts and theories of physics grow.

T: Excuse me, is the Pauli-Jung correspondence published? I have seen interesting remarks about it in your articles but I did not know that it is now available in printed form.

P: It was published in 1992 by Springer-Verlag (*Pauli & Jung* 1992). The editor is an elderly psychologist, C.A. Meier, a friend of Jung and Pauli, and he was assisted by Pauli's last assistant C.P. Enz, now professor emeritus of theoretical physics at the University of Geneva.

Pauli was greatly interested in the unconscious background of physics research. He found the concept of archetype very important. He often spoke of the 'ordering and regulating' [das Ordnende und Regulierende] which is the basis of all knowledge, both in psychology and in the natural sciences. Thus, archetypes also bind the material and the psychical inseparably to each other. From this basis he tried to approach the psychophysical problem, which he called the most important problem of our time.

Pauli considered archetypes as being even more abstract than did Jung in his last years. Mathematics is, in Pauli's view, the clearest evidence of the archetypes. In group theory, in particular, we meet general structures – for example, symmetries – whose manifestations we see everywhere: in crystals, in the charming forms of flowers as well as in the structures of organisms more generally, and also in products of the arts which gain much of their beauty from certain symmetries. But symmetries and group theory also find abstract expressions in natural laws. Theoretical physics can essentially be understood as a description of symmetries inherent in nature and in natural laws.

Here I would like to speak of natural numbers or integers which Jung as well considered to be profound expressions of archetypes. I mean the endless series of numbers 1, 2, 3, ... Here we may not confine ourselves to the arithmetic properties of integers. We are now mostly interested in number mysticism. It helps us to shape the unconscious background from which the arithmetical properties of integers are born. As I mentioned before, number mysticism seems to be related to the trinitarian theology – you can certainly say more about this relation. One has to shape the basis of existence and human knowledge, and if we believe that God is the origin of existence, we are here attempting to learn something about God from the basic structure of the Creation.

T: As a Lutheran theologian, I must, however, state that God is for me an object of belief, not of reasoning. What I have learned about the Holy Trinity, I have learned from the Bible and from the articles of faith, not from any number mysticism which for me seems to be related to superstition.

P: It was good that you mentioned this because here our attitudes seem to be clearly different. If I am told that the condition for my salvation is that I believe everything that theologians teach on the basis of the Bible and the articles of faith, my critical instinct will rise in rebellion. Theologians have taught matters which later have been shown to be clearly untenable. I cannot blindly believe anything; instead, I try to doubt everything before I accept something new as true. For me, becoming familiar with number mysticism has been a key to the mystery of the Trinity. This mystery seems, in fact, to be an almost insuperable barrier to the acceptance of Christianity among the intellectuals of today. By the way, I would like to remind you that according to Pavel Florensky, faith is complete only when it is also accepted by critical thought. His maxim was: "Intelligo ut credam".

That the doctrine of the Trinity can become an obstacle to educated people today is a consequence of the 'requirements of scientific exactness' which philosophers used to emphasize. These requirements are good in science, and sound criticism is always necessary. However, one must remember that there are also irrational dimensions in reality, and the requirements of rationality are not appropriate when describing them. Religion speaks mostly of those irrational dimensions, and if we wish to speak of such matters at all, this is not possible in the same way as in the exact sciences. The trinitarian theology is a good example. For me, number mysticism has opened the way into it. I find it to be a very profound description of the basic features of existence – as far as one is willing to comprehend in some way the irrational origin of science. When I try to describe these matters, I explicitly start from number mysticism – as I understand it.

T: The creeds which express the doctrine of the Trinity were a result of a long development and maturing work. For example, the Nicene-Constantinopolitan creed was finally accepted only in 451. In this form it can be called ecumenical: It is used by practically all Christian churches (if we omit the famous 'filioque' controversy which divided the Christian world five hundred years later).

P: During this maturing process, number mysticism – especially in the form of Pythagorean and Platonic philosophy – probably had an influence on the final form of the idea of the Trinity. I understand that number mysticism contains in a concentrated form certain fundamental features in the unconscious functioning of the psyche, and simultaneously, these features reflect the basic aspects of existence; they are important, of course, if we try to get an image of God. One should remember that the idea of trinity also appears, in various shades, in some other religions. This is an ecumenical matter, in a very profound sense. For this very reason one should not belittle number mysticism when speaking of the doctrine of the Trinity.

I shall repeat the main ideas. The number 'one' does not have any sense alone because the idea of 'one' already contains an idea of something else. Thus, 'one' necessarily presupposes number 'two' as well: there is something that can be distinguished from 'one'.

But if there are two things distinguished from each other, the question arises of what their relation is to each other. This relation is a new logical object, and this is the idea of 'three'. These three things, 'one', 'two', and 'three', form a certain whole, a unity, which in a qualitative sense contains everything characteristic of the integers in general.

From the point of view of the interpretation of the world of integers, it is important what the nature of the relation is which number 'three' represents. It unites the numbers 'one', 'two', and 'three' into a unity. This unity is called *trinity*. In the Christian doctrine of the Trinity, 'one' is the Father, 'two' the Son, and 'three' the Holy Ghost. Most delicate is the nature of the Holy Ghost. The part of the Nicene creed which concerns the Holy Ghost got its form last, and there have been and still are different views concerning the Holy Ghost.

In Plato's *Timaeus*, 'three' is interpreted as a mathematical relation, which Plato considers the most beautiful way to relate two things to each other. This is characteristic of Plato's rationalism. According to St. Thomas Aquinas, the basic nature of the Holy Ghost is love. It is the uniting force of the Godhead – although the Godhead contains three Persons or hypostases. It is essential that all three Persons are of the *same substance* (essence). This is the *homoousian* idea characteristic of the doctrine of Trinity.

T: I am no expert on trinitarian theology, but I think that this description is quite acceptable. Love is, of course, the basic feature of the Christian image of God. Also important is the superhuman intelligence and power which can be seen in God's creative activity, described by such attributes as 'omniscient' and 'almighty'. These are aspects of God's basic nature which, according to the homoousian idea, concern all three persons.

In the splitting of the Church into an Eastern and a Western part, different views concerning the Holy Ghost formed a severe obstacle. According to the Orthodox conception, the Holy Ghost is a 'spirit' or 'pneuma' emanating from the Father, while in the Roman Catholic formulation, the Holy Ghost emanates from the Father and the Son (Filioque = and from the Son).

I think we can use number mysticism in the way you have proposed, pointing out that the basic nature of the Holy Ghost and, thus, of the Godhead is love. It is primarily the force which binds the Godhead to a unity. However, it is also important to point out that all Persons (hypostases) have the same substance. Thus, love belongs to the nature of all Persons, not only to the Holy Ghost. Neither should one forget that truth also belongs to the nature of the Godhead, and it also unites the three hypostases in the sense of the homoousian idea.

The Unity of the Spirit

P: Thus we have met the important question of the role of the 'third' as the uniting element. In Christian thought, the uniting force is, primarily, love. You quite correctly pointed out, however, that one should not forget truth, which also belongs to the nature of God. Truth means correct comprehension. Both truth and love presuppose that they have an object, and this leads to creation. This makes the genesis of the Creation understandable. Truth and love also form the basis for the relation between God and his Creation, but I have to postpone such questions until we discuss the *quaternity*. In any case, *the uniting element is the 'third'*.

Using the Christian terminology, we can say that the Holy Ghost unites the congregation and is also the basis for man's communion with God. The Creation is outside of the Trinity, and as the object of God's love and comprehension, it contains elements that distinguish it from God. Therefore, the relation between God and the Creation is problematic, and the Holy Ghost is needed as the uniting force. But let us postpone these problems a little until we speak of the quaternity.

T: I must state that you describe well the Christian doctrine of the Trinity. Perhaps number mysticism has had a certain influence on the final formulation of this doctrine. If you have learned these things primarily from number mysticism, instead of from the Bible and the articles of faith, I find it strange that you seem to understand these very abstract matters so well.

P: The idea of the action of the Holy Ghost is personally quieting – as I understand it. It looks as if theologians would somehow like to 'bar me': They don't like the problems which are obstacles for me, as a disciple of the exact sciences. There *are* problems if we wish to find harmony between religion and the picture of the world constructed using the methods of empirical science. For theologians these problems seem to be unnecessary philosophizing and even insulting to true religiosity. It is comforting that the Holy Ghost can open a way even for a free thinker of this kind.

Orthodox Archbishop Paavali, before his last Christmas, unexpectedly wished to visit our home during that Christmas time which he spent in the Helsinki area. His visit left a very strongly uniting feeling, something like that described in hymn 454 in the hymn book of the Finnish Lutheran Church. Afterwards, we sent him this hymn as a message.

I believe that the Holy Ghost does its uniting work both among the Christian churches and outside of them. For me it is difficult to understand those who want to separate the 'true believers' into their own group of the saved and classify the others as heretical, on the basis of different characteristics. The orthodox archbishop did not leave such an impression.

The emphasis on orthodoxy always leads to the condemning of the others. It is the origin of separatism or even of religious wars. "Judge not, that ye be not judged" was declared in the Sermon on the Mount. The Holy Ghost unites; separatism has its origin in the Devil, the binarius.

In order to illuminate the nature of number mysticism, I would like to discuss another viewpoint associated with psychology. Let us think that 'one' means me: a subject who aims at understanding what he experiences – tries to give shape (gestalt) to his experiences. A presupposition for shaping the world is that 'I' divide my consciousness into two parts; these parts are 'I' (ego) which acts as the active subject of shaping, and the 'world' which is the object of the shaping process. Here are these 'one' and 'two' which are the presupposition for comprehending reality – in the sense that I could speak of 'truth'. It is also clear that there is some relation between them ('I' and the 'world') because otherwise the knowledge of the world would not be possible for me. But it is extremely difficult to explain the nature of this relation. In any case, the subject and the object must in some sense form a whole, be in 'interaction' with each other, in order for the subject to gain some knowledge of the object.

In quantum physics, the question concerning this interaction between the object and the subject has been found to be extremely difficult. According to Bohr, the measuring instrument (which represents the subject) and the atomic object form an indivisible whole in an observation; it is senseless to speak of the details of this interaction. The very fact that this 'interaction' cannot be described in a rational way (using causal laws) is the heart of the 'epistemological lesson' of quantum theory. There is an analogy between the holistic feature of quantum theory and the uniting nature of the Third Person which is perhaps not without interest.

In order to truly understand this analogy, it is necessary, however, to take one more step in number mysticism: the decisive step from 'three' to 'four'. We should now proceed from *trinity to quaternity*.

T: OK, now we are coming to the really dangerous questions, but this was your wish. Certainly we are now arriving at a field which is completely unknown to me.

Unfortunately, I have been called to an unexpected meeting, and I must leave. I propose that we continue this discussion another time.

P: Here, in one week?

T: That's fine. We shall meet here a week today.

12. The Problem of the Fourth

Trinity and Quaternity

Theologian: Here we are again. You wished to discuss the idea of 'quaternity', a quite strange concept for me, and you said that it also opens a perspective to creation. To be frank, I am a little skeptical.

Physicist: OK, the question is how does one proceed in number mysticism from number 'three' forward, and what is the nature of number 'four', as compared with the trinity formed by 'one', 'two', and 'three'? Because a trinity forms a unity, a whole, it can be compared with the number 'one'; analogously it cannot exist alone. In the Christian Godhead, the main attributes are truth and love, and these presuppose an object. This object of a trinity is 'four'. Its existence is not as imperative as the existence of 'two' for the number 'one'. Rather, one can say that this is a question of an *analogy*: In the same way as the number 'two' is necessary for the number 'one', the unity which the trinity forms must have an object, and this is the number 'four'.

It is exactly this step which represents the idea of *creation*. According to the inner dynamics of the trinity, this unity creates an object for itself. The trinity and 'four', then, form a whole which is called a *quaternity*. However, the basic nature of this whole is essentially different from that of the trinity.

This presupposes some relation between 'four' and the trinity which makes them a whole. This relation must again correspond to the inner dynamics of the trinity, which is the basis of creation. The relation between 'four' and the trinity must, in principle, be the same as the relation between 'one' and 'two' – which makes 'one' and 'two' a unity.

When thinking of the Christian doctrine of the Trinity, we must say that 'four' represents God's Creation, the created world. Its relation to the Godhead is analogous to the relation which makes the Trinity a unity. The main attributes of this relation are truth and love. There is, however, an essential difference between the quaternity and the trinity: The trinity is a necessary consequence of the existence of 'one', while the existence of the quaternity is only based on analogy, reflecting the inner dynamics of the trinity. The idea of creation, the proceeding from the trinity to 'four', contains an element of freedom which is something different from the necessary unity of the trinity.

T: Is it not also possible to arrive at the problems of morality on this basis?

P: Quite right. This freedom contains the *problem of evil* which makes it impossible for many people to believe in the existence of God: If God exists and is almighty and omniscient, how is it possible that there is so much evil, terrible evil, in his creation?

I would like to compare the trinity with a completely rational system. The eternal process in which 'two' is born from 'one' and united to it by 'three' can be compared with a logical necessity. Since 'two' is, in a way, a counter-being to 'one', this would create a tension between these beings if this were not compensated by the binding effect of 'three', which makes the trinity a unity. Analogously, 'four' is a counter-being to the trinity, but the bond which forms the quaternity, is not as absolute as the bond which makes the trinity a whole. The difference between these bonds is the freedom characteristic of creation, and this brings in irrationality into the quaternity. Beside rationality the quaternity contains irrationality, while the trinity is purely rational – in this simile.

If we apply this to the Christian doctrine of the Trinity, the quaternity also contains the real opposite of the Godhead: the Devil. Because of the freedom which belongs to creation, this real counter-force to God also belongs to the created world: It distinguishes the Creation from its Creator. This helps us to understand the nature of the irrationality which necessarily belongs to the world in which we are living. The freedom that is characteristic of the act of creation makes the created world a battle field where God and His counter-force – love and hate, truth and untruth, good and evil – continuously fight against each other. Creation is a continuous process, and this fighting belongs to its very nature. In it our faith is proved, whether it is directed towards God or away from him.

T: Have you, indeed, arrived at these ideas from quantum theory?

P: I would claim so. Decisive, of course, is the decline of the deterministic picture of the world. It brings freedom into phenomena, this fighting, but also the possibility to see teleology in creation. This brings meaning into existence. I felt this very clearly during an autumn more than twenty years ago when I had a slight heart infarct which caused a change in life in many respects. My mother died in the same autumn, and I was with her much of the time at the end of her life. I then had a strong feeling that God is present in everything that takes place. For me, it was the end of scientism, and I began to see the world in another light. As I simultaneously gave lectures in the philosophy of modern physics, these feelings were closely connected with the philosophical ideas of quantum theory. I began to understand that even scientific research is based on some belief which gives direction to the endeavor. It is the direction of this belief which is decisive.

When, some years later, I encountered Pauli's letters, a new world was revealed to me. What I have related here is an attempt to formulate in an understandable way views which Pauli has created in me.

T: Can you say more exactly how quantum theory has given you these kinds of ideas – in addition to its showing that the deterministic picture of the world is untenable?

P: It has been very important to see that we need another 'element of reality' in the interpretation of atomic theory, in addition to matter. I mean consciousness and free will – let us say *psyche*. The 'paradoxes' of quantum theory disappear if we understand that the human psyche has an essential role in the interpretation of experimental results. In fact, the microphysical object cannot be separated from the observer's psyche in an observation. This follows from the finite value ($\neq 0$) of the quantum of action (Planck's constant) which makes an observation process an indivisible whole. The same thing can also be described with the aid of the uncertainty relations of Heisenberg. These are basic features in the original Copenhagen interpretation (but today often incorrectly presented).

This has resulted in an ontology corresponding to the Unus Mundus of Pauli and Jung. When describing reality, we need physical and psychical phenomena as its complementary expressions. In the background, we can feel that we see something of the transcendent reality itself which I would like to call the *world of spirit*, as I said last time.

It is astonishing how strongly the materialistic basic view still influences physics research today. Almost everyone tries to describe atomic phenomena in such a way that the psyche is not needed in the description. In addition, such attempts are called 'realistic', although this means that facts which show that traditional realism is not compatible with quantum mechanics are neglected. People do not wish to accept the introduction of psychic aspects into physics.

T: This does not explain how you have come to number mysticism.

P: Because of my interest in Pauli and Jung. For them, the unconscious processes of the psyche were important, especially the role of archetypes. Pauli found integers to have an archetypal nature, both in shaping a picture of the outer world and in the functioning of our psyche in general. The downright mysterious force of mathematics in physics research strongly influenced Pauli's philosophical views. Eugene Wigner has written an article about "the unreasonable effectiveness of mathematics in the natural sciences." (*Wigner* 1960.) Mathematics leads a scientist like a higher, supernatural force. This mystical experience gave Pauli reason to speak of a *cosmic order* behind the distinction physical-psychic. (*Pauli* 1994, e.g., p. 34.)

The Problem of Evil

P: There remains one aspect of the quaternity that we should discuss in more detail. I mean its connection with the problem of evil, which was only briefly touched above. From the point of view of religion, it is an important question, isn't it? In the atheistic propaganda, the absurdity of the belief in a God of love is strongly emphasized.

I have the feeling that Christian churches are inclined to belittle this problem. The goodness and love of God is accentuated too one-sidedly. In the Old Testament, we meet a quite different God: a terrifying, violent, capricious ruler who rewards fathers' deeds down to the third and fourth generation, independently of how these ill-fated people live and try to worship and honor him. The Creator is more like a natural force than a human being – and this picture corresponds well to the picture the natural sciences give of this universe.

Is it possible that neo-Platonism has in this respect had a distorting influence on the image of God in Christianity? It has strongly influenced Saint Augustine's thought. For neo-Platonists evil is just a lack of good, *privatio boni*; they scarcely knew of active evil. In moral questions, Greek philosophy after Socrates was inclined to an unrealistic attitude, believing that the knowledge of moral questions implies moral behavior. Perhaps the Pythagoreans and Heraclitus understood these matters more realistically.

T: The Christian theology has certainly been influenced by neo-Platonism in this respect. However, the problem of evil was also difficult for Luther. He speaks of the left-hand deeds of God in addition to his right-hand deeds. If we consider what has happened in the unfortunate former Yugoslavia, it is really difficult to agree that the God of love has allowed all this to happen if he is almighty.

P: In this connection Pauli has referred to a Platonist, Scotus Eriugena, who lived in the 10th century, and to his interesting description of the three persons of God. I have mentioned this in different connections, but perhaps we should think about his ideas even today because this associates the problem of evil explicitly with the 'problem of the fourth' which we are now discussing.

The lower part of the frontispiece to Part III originates from Scotus Eriugena. Its structure is the same as in the upper figure on the same page; the latter is used to illustrate the connection between the four elements and two basic opposites. When describing the properties of the four elements, these opposites are cold / warm and dry / wet.

The opposites presupposed in the figure by Scotus Eriugena were: *that which creates / that which does not create* and *that which is created / that which is not created*. If one has to describe the persons of the Creator, this seems to be a suitable starting point. According to Scotus Eriugena, the three persons or hypostases of God can be described as follows:

1. *That which creates and is not created.* This is God the Father, who is the

origin of creation and, according to Scotus, its aim as well, because existence means a circular change where all that is created in the end returns to God.

2. *That which creates and is created.* This is God the Son, through whom everything is created. (In fact, this characterization of God the Son does not quite correspond to the ecumenical creed according to which the Son is born, not created.)

3. *That which is created and does not create.* This is the created world. Scotus Eriugena associates this with the Holy Ghost in which form God is present in the created world.

Interesting here is point 4 on the circle. It does not represent any person in the Trinity. It should be characterized by "that which is not created and does not create." Scotus interprets it with *deification,* which means returning to God, to the origin. This is described by the statement: "Finis enim totius motus est principium sui; non enim alio fine terminatur nisi suo principio a qui incipit moveri." (The aim of all motion is namely its origin; it is namely not limited by any other aim than its origin, from which motion begins.) Thus, there is no person corresponding to point 4, but it represents the return to the origin of the cycle.

This is characteristic of Greek thought, where everything returns again and again to the original state in an eternal cycle. The medieval alchemists, who anticipated the birth of modern science, had a different view of the happenings in the world. This would correspond to a change in Scotus Eriugena's figure, where point 4 would be interpreted as the Devil! The Trinity is then replaced by a Quaternity. I am inclined to say that this kind of image of God corresponds better than the trinitarian one to the order dominating in the created world.

T: I have seen this story in one of your books. The Devil is real and cannot be neglected. In Christian parlance we speak of a fallen angel. The idea which you presented earlier – that the created world must necessarily contain something that distinguishes it from its Creator and that, therefore, creation in this sense also brings forth the evil – is interesting; it makes the existence of the evil in some way understandable. In Luther's language, point 4 would correspond to the left-hand deeds of God.

I don't think we can avoid admitting that there is irrationality in this world, something that human reason never can comprehend. We must just accept this fact, and there our belief is proved, in our meeting this irrationality. In the Book of Job there is a shocking description of these problems and the left-hand deeds of God. (By the way, they are caused by Satan's advice.) Without this irrationality, evolution would not be possible. In fact, the whole existence of the organic nature is based on cruelty, the struggle for existence which continually goes on in nature. Everyone can feel the nature of this struggle in his/her heart.

I would like to emphasize, however, that we seek from religion strength for life, for this struggle for existence. Therefore the Christian faith is based on the Trinity, and the fourth person does not belong to God. The more we can live as if the fourth person does not exist at all, the closer we are to God. Therefore, the Comforter, the Holy Ghost, has been sent to us, and he glorifies the true nature of God to us; in our previous discussion we have spoken of this.

Christianity also has the vision of the Last Judgment, where everything that distinguishes the Creation from God will be burnt in the fires of hell. In that respect, Scotus Eriugena's view corresponds to the basic Christian ideas – although it does not correspond to the present situation in the created world.

P: I can readily agree with these views. Religion must give us strength to live. It should glorify to us the very nature of God and help us to fight against that which separates the world from God. Therefore the Christian God of love makes a stronger appeal to me than the Jewish Yahveh.

Quaternity in Science

T: I am still thinking of one question concerning quaternity. I remember that Pauli studied the influence of archetypes on Kepler's scientific work in particular. In that connection he found a controversy between Kepler and the English physician and alchemist Robert Fludd. I think Pauli describes Kepler as a typical representative of the trinitarian attitude and Fludd as a quaternarian thinker. What did he really mean by the trinitarian and, correspondingly, by the quaternarian attitude in this connection?

P: This is, indeed, a difficult question for me. I would like to say that the trinitarian attitude is now the dominating attitude in science, where only those matters are considered real that can be described by using a logically correct language and by continually submitting the truth of statements to empirical tests. Practically, this means that science is based on a deterministic picture of the world, where all events have natural causes which can, in principle, be found with the aid of scientific methods.

Statistical causality breaks this tradition: one cannot find 'natural causes' which completely explain phenomena. The trinitarian attitude presupposes, however, that the formal requirements of science must be kept in force, and therefore people refuse to discuss matters which cannot be described by using an exact, scientific language.

The quaternarian attitude, on the other hand, accepts intuitive insights, and then it is possible to also speak of explanations that are metaphysical by nature and perhaps downright 'supernatural'. The controversy between Kepler and Fludd which concerned the explanation of the planetary motion is an example of a clash between these different attitudes. Fludd writes that Kepler is only interested in the tail of beings, in the observable phenomena. He, Fludd, is searching instead for primary causes, and therefore he is embracing the head. Kepler then replies: "I hold the tail but I hold it in my hand; *you may grasp the head* mentally, though only, I fear, in your dreams. I am content with the effects, that is, the movement of the planets."

Modern natural science has followed the way laid down by Kepler, and undeniably with good results. Atomic physics, however, has brought forth problems which unveil the limitations of the present empirical methods. Pauli came to admit that rational science must accept intuitive insights on a par with it; otherwise, we arbitrarily omit essential facts. These cannot be described other than by using metaphors and analogies, but these are needed for judging questions concerning the direction of science which are of current interest. The role of science in general cultural evolution, perhaps especially the relation between science and religion, must be re-evaluated. The quaternarian attitude means that we don't neglect such more general obligations when judging the direction and methods of science.

This attitude has concrete implications even for the direction of research within physics itself. One can claim, with weighty reasons, that *quantum mechanics leads to a psychophysical ontology.* Then the problems of fundamental research are completely changed. Simultaneously, one needs, for example, re-judgement of the basic philosophy in present cosmological models and present theories of evolution including biology. The quaternarian basic view may demand radical changes in various fields of empirical research.

This does not mean that the rational methods of science will lose their value or that they have to be made less exact. One should only take into account the limitations of these methods. It is also important that a philosophical analysis of the basic concepts is not excluded.

However, we probably cannot progress further in these difficult questions today. I am very grateful that I had this opportunity to discuss these matters with you. They definitely require more clarification now. We are at a turning point, and many things will be changed. I think that these questions which we have discussed belong to the essential problems of today.

I must add that I feel the reality of hell very strongly. The development which has guided our cultural endeavors seems to be leading us to a catastrophe. It is not directed toward God. On the contrary, love has been replaced by selfishness, truth by illusions created by an overemphasis on reason. We are at a turning point, and I am not sure that we shall be able to find the right direction.

T: I wish to thank you for the initiative of these discussions. Certainly these kinds of dialogues produce new ideas. Let them grow in our quiet moments. I believe that the Comforter will help us to find the way. And give us strength to fight against the temptations of the Devil. God be with you!

IV

The Outline of Reality

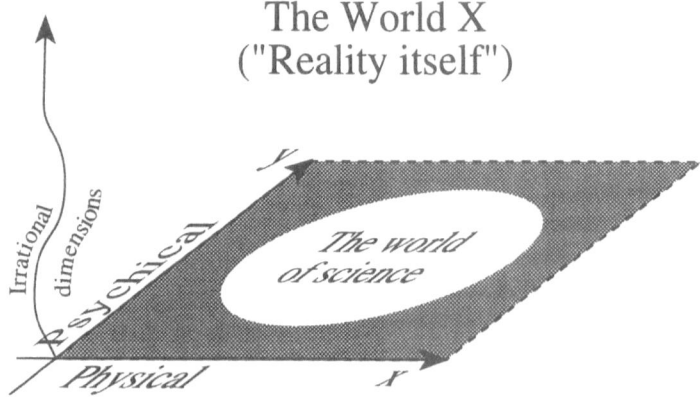

The World X
("Reality itself")

Transcendent reality

The unus mundus of Jung and Pauli or the 'world X' of Pauli is transcendent by nature, i.e., it is not reachable by human thought. It is represented by the three-dimensional space of the figure. The world described by the rational methods of science is a plane intersection of this space. The third dimension visualizes the irrationality of reality.

13. The Psychophysical World

The Problem of Realism

Let us recapitulate the birth of quantum mechanics, with some additional remarks.

One day in the year 1900 the German physicist Max Planck was walking with his son in a Berlin park. He told the boy that perhaps he had found a law of nature which could be compared with Newton's laws. This is the story his son later told. Max Planck had found the quantum, the idea that there is a smallest amount of energy for each kind of radiation – an amount which cannot be subdivided in any way.

This was the beginning of a *discrete physics* or the physics of indivisible wholes. All the laws of physics had thus far presupposed continuity: that each quantity could always be subdivided into ever smaller quantities of the same kind. Planck found that this is not true with respect to radiation energy. Radiation is always emitted as certain 'energy packets' which cannot be subdivided into smaller amounts of energy. This was in contradiction to the basic *axiom of continuity* in classical physics – as we now call the physics which existed before the discovery of the quantum.

From this beginning, quantum physics has developed into the most revolutionary theory of modern physics. Quantum has been a very reliable guide, and it has helped physicists to understand finer and finer structures in matter, first in atomic physics, then in the investigation of the atomic nucleus and of the elementary particles. Since the discovery of Newton's laws 300 years ago, there has been no theory in physics more successful than quantum theory.

There is, however, something mysterious about this theory. Its strongest development occurred between 1924 and 1927 when the so-called *quantum mechanics* was developed in close collaboration between experimental physicists and theorists. The idea of the indivisible quantum was then reconciled with the continuum ideas of classical physics, and this mixing of discreteness and continuity was the key which opened the door to the secrets of the microworld. Bohr, Heisenberg, Pauli, Schrödinger, Born, Dirac and many other theorists participated in this strong development which opened a new world for physics. But when the new theory was finished, physicists found that they did not understand what they had done. The *Copenhagen interpretation* of quantum physics was presented by Niels Bohr in 1927 and was generally accepted as the language for describing microphysics, but very few physicists really understood the radical nature of Bohr's philosophy. Perhaps not even Bohr himself.

When a new generation of physicists occupied university chairs in physics after the Second World War, a very strange situation was realized. This younger generation knew the enormous strength of quantum physics very well but they found that they did not really understand it. The result has been a pragmatic attitude among physicists: they don't discuss philosophical problems concerning the foundations of quantum theory but they use it. And it works well according to the recipes Bohr provided.

There is, however, a growing group of physicists and philosophers who have found a regular mine of information – and of publications – in the foundations of quantum theory. This interest in the philosophical problems of atomic theory has greatly increased during the last twenty years, after the so-called EPR-experiments showed that *Einstein's realism leads to contradictions with experimental facts*. This shows that there are real problems concerning realism in the atomic world.

The difficulty is that the same atomic 'objects' seem have both particle properties and wave properties, depending on the experiment which is chosen for their investigation. These experimental arrangements are mutually exclusive. This is the famous wave-particle dualism characteristic of the microworld.

What kind of object is an electron? Is it a particle? No, because it has wave properties in certain experiments. It seems to be an object which has *both* particle *and* wave properties – but this would mean that it is in contradiction to itself. This is the difficulty concerning the microworld. Its objects cannot be *either-or*, they should be *both-and*, and normal logic does not allow this. The microworld seems to have self-contradictory properties. We can no longer claim that reality cannot be in contradiction with itself. *There seems to be something irrational in reality.*

Bohr could not draw this conclusion from his beloved quantum mechanics, and therefore he did not wish to speak at all of reality or realism. However, he adopted the well-known Taoist Yin-Yang symbol (Fig. 1) for his coat-of-arms in the Danish Elephant Order, and he often described the paradoxes of reality by saying that according to the normal logic the opposite of a true statement is certainly not true, but *the opposite of a deep truth can very well be another deep truth.*

Figure 1. The Taoist Yin–Yang symbol for reality composed of two opposites which continually give place to each other.

Wolfgang Pauli has clearly discussed the ontological implications of quantum theory. He boldly states that we must acknowledge the *irrationality of reality*. This means that reality cannot be described by using the normal 'either-or' logic. We must accept that there are features in reality which presuppose a *'both-and'* logic. *Reality has self-contradictory properties.*

This is an implication of atomic physics which people do not wish to accept. Both physicists and philosophers have tried to avoid the irrationality of the atomic world in all possible ways, and the view which Pauli presents has been absolutely excluded so far. After having studied Pauli's thought for about 20 years, I must state that Pauli has really understood the lesson which atomic physics has given us. It concerns not only atomic physics, but also empirical knowledge in general. I shall try to illustrate this by describing Pauli's view a little more specifically.

Pauli's Conception of Reality

Pauli had certain psychic difficulties when he was about 30 years old. He then came into contact with the Swiss psychiatrist C.G. Jung, and this contact continued until Pauli's death in 1958, not in the form of medical consultations but as a very interesting scientific contact which until recently was not well known. Now this collaboration between these two men of genius seems to be arousing increasing attention. Together they published *Naturerklärung und Psyche* in 1952; an English translation *The Interpretation of Nature and the Psyche* was published in 1955 (see *Jung & Pauli* 1952). This volume contains two formally independent articles. Jung's article is called "Synchronicity. An Acausal Connecting Principle," and Pauli's article has the title "The Influence of Archetypal Ideas on the Scientific Theories of Kepler." The common denominator in these articles is the illumination of the functioning of the unconscious and especially the role of the archetypes. This is the main theme in the "collaboration" between these two scientists. It was not a planned collaboration but just a continuing contact for discussion of questions which both of them considered extraordinarily important.

Jung and Pauli both lived in the region south of Zurich, Jung in Küsnacht and Pauli in Zollikon. In addition to personal conversations, they frequently communicated by letter, and this correspondence was published in 1992: *Wolfgang Pauli und C.G. Jung. Ein Briefwechsel 1932–1958*, edited by C.A. Meier. This is a very interesting volume, and it especially illuminates the development of Pauli's thought where the ontological problems of quantum mechanics are continually the center of interest. Fundamental for Pauli was the *general order in the cosmos* which finds its expression both in the laws of nature and in the functioning of our psyche. Pauli's interest in archetypes was primarily based on their appearance in the cosmic order of the world, but he also became more and more interested in their role in the functioning of our unconscious psyche. I have the feeling that this is a view which will help Western culture to find new unifying ideas, in place of the continuing specialization which is splitting it in a dangerous way.

Pauli combines the ontological paradoxes of atomic physics with psychology. In fact, this is natural because physics is based on observations (measurements), and observations are psychic phenomena. In the analysis of the basic problems of quantum theory, the description of observations has been the central issue. In general, physicists avoid psychological questions, however, and try to describe measurements as purely physical phenomena. Pauli had a different view, and, in fact, he presents the implications of the Copenhagen interpretation in the most natural way. For Pauli, only physics and psychology together can describe reality in a reliable way. Physics can only describe the rational features of reality because its theories are bound to the rules of the two-valued 'either-or' logic. We must include psyche in our conception of reality in order to understand the irrational aspects of reality.

For the founding fathers of quantum mechanics, it became increasingly clear that it is not possible to describe the atomic world in the sense of normal realism. Especially in the 1950s, they often stated that the quantum mechanical description of atomic events aims not at a description of the atomic world but just at a description of our knowledge of this world. This presupposes that there is, in reality, something in addition to the material world that our theory describes: something 'that knows' – consciousness. In the interpretation of quantum theory, consciousness is necessary, at least implicitly. Pauli states it explicitly. But in addition to consciousness, the *unconscious* plays an essential role in Pauli's thought. I mean the Jungian unconscious, not the subconsciousness which is often used in Freudian psychology. This I shall try to illuminate with the aid of some pictures (Figs. 2 and 3).

The smaller circle in Fig. 2 represents the consciousness of a person. Around the circle of the consciousness there is another bigger circle. The area between these two circles represents the *subconsciousness* in the Freudian sense. Beyond these circles is the outer world which is the object of physics and other natural sciences. Everything that comes from the outer world into the consciousness comes through the subconsciousness, because the impulses coming from the outer world give rise to unconscious processes which shape them into a form which we perceive as certain conscious things. Thus, the subconsciousness is a mediator between the outer world and the consciousness. Of course it also has other tasks which Freud in particular emphasized, but we do not need to discuss these things in more detail. In any case, the subconsciousness is a realm of unconscious instincts, and, as if it were behind a curtain, it strongly influences what takes place in the consciousness.

I try to describe Jung's unconscious in Fig. 3. The circle again represents the consciousness, but now it is surrounded by *the unconscious*, which is infinitely deep: It does not have any outer limit. Now where is the outer world which is the object of physics? I do not know. Everything that comes into the consciousness comes from the infinitely deep unconscious. This situation describes very well the

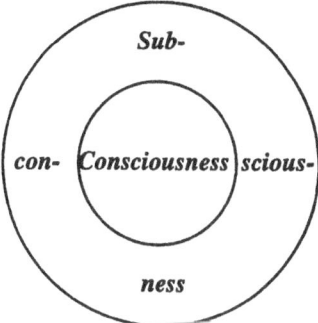

Figure 2. In Freud's depth psychology the unconscious / subconsciousness is like a reservoir into which matters can be removed (forgotten or rejected) from the consciousness for a shorter or longer time – or for ever.

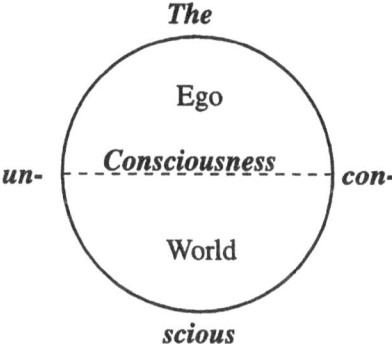

Figure 3. The unconscious in Jung's psychology is infinitely deep containing everything that can come into the consciousness, including the transcendent 'outer world'. Cf. open realism of d'Espagnat (*d'Espagnat* 1993).

problems of reality in quantum mechanics. In fact, we cannot prove, in a strictly scientific sense, that there is any outer world. In principle, we just believe that there is something behind the pictures we have in our consciousness. These pictures of the consciousness we *project into the 'outer world'*, and we strongly believe that this outer world really exists. This belief is strengthened by the fact that we can predict many things on the basis of our conscious perceptions and of our knowledge, and especially on the basis of science. *But in principle the existence of the outer world is just a belief.* It is the basic belief of empirical science because without this belief there would not be any sense in scientific work.

It is important to understand this situation. In January 1955, some months before his death, Einstein wrote to M. Laserna:

It is basic for physics that one assumes a real world existing independently from any act of perception. But this we do not know. We take it only as a programme in our scientific endeavors. This programme is, of course, prescientific and our ordinary language is already based on it. (Translation according to *Fine* 1986, p. 95.)

If you wish to localize the outer world somewhere in our Fig. 3, you can say that it is in the infinity. It is good, in some respects, to imagine that our plane is the 'function-theoretical plane' where there is just one point at infinity: In all directions you reach in infinity at the same point. This is characteristic of the 'function-theoretical plane.' This *point of infinity* represents the outer world in Fig. 3. Thus, we imagine that all impulses coming 'from the outer world' into our consciousness come from this point of infinity. But these impulses acquire the structure in which we perceive the outer world through the 'shaping' (gestalting) processes in the unconscious.

The idea of *independent reality* which is absolutely objective and independent of any observers and observations is outside the realm of empirical science because empirical science is based on observations. Therefore the idea of independent reality is just a belief, although it is a belief which is basic for the whole endeavor.

Bernard d'Espagnat has very clearly analyzed the concept of reality, which becomes problematic in quantum theory. He calls the new form of realism, which quantum mechanics seems to presuppose, *open realism* (*d'Espagnat* 1993). He emphasizes that it is necessary to speak of an independent reality but, on the other hand, that we cannot describe it in the sense of traditional realism. We can only have certain statistical knowledge of it: We can construct a reliable theory which gives statistical predictions for atomic events, but *only statistical* (probabilistic) predictions. We have certain criteria for the reliability of our theories in this respect, and therefore we can say that we get a 'reliable' picture of reality, although we cannot reach the reality itself, not even asymptotically. This picture of reality, which is formed according to scientific principles, d'Espagnat calls *empirical reality*. It is a changing picture because science develops, and we cannot claim that it describes independent reality itself. But for many purposes this picture – this empirical reality – is useful. It is based on our conscious perceptions and theories constructed by the human consciousness. It has certain value for us, as we know from experience, but independent reality remains forever 'behind a veil'. Therefore d'Espagnat speaks of *veiled reality*. Independent reality is the 'mysterious' point of infinity in Fig. 3.

Here I have tried to describe Pauli's conception of reality. It was a result of the collaboration with Jung, and depth-psychology plays an essential role in it. d'Espagnat came to his conception of veiled reality in a quite different way. His conception of reality is, however, quite the same as that of Pauli. They merely use different words: d'Espagnat speaks of veiled reality, Pauli speaks of the irrationality of reality. According to my understanding, this kind of idea of reality is a necessary implication of quantum mechanics, if its Copenhagen interpretation is accepted. However, only a few physicists accept this kind of realism so far, because normally physicists are not willing to mix physics with psychology. But I am sure that Pauli's conception of reality opens a perspective which is important for the future development of science in general – not least for the development of physics.

Why does d'Espagnat call the new kind of realism, based on this conception of reality, open realism? It is because we must leave open the properties of independent reality. We just state that there is an independent reality. But we do not make any assumptions about its properties. Especially, *we do not presuppose that it is describable in the scientific sense.* Einstein presupposed that reality is comprehensible and even mathematically describable. The EPR experiments have shown, however, that this assumption led him to wrong predictions, at least if we also accept the theory of relativity. Our physical experience in general seems to speak against the idea that we could, in the future, learn to describe independent reality. Therefore the metaphor of 'veiled reality' seems to be good.

The Psychophysical Problem

Since the 17th century, Western thought has been strongly influenced by *Cartesian dualism*. On this basis, the 'spiritual world' has been excluded from the natural sciences because the 'world of matter' is considered, in principle, independent of the 'world of spirit.' This leads to a materialistic attitude which indeed dominates in the natural sciences.

Quantum mechanics is a turning point. The Copenhagen interpretation of quantum mechanics implies that we introduce consciousness into physics. Quantum mechanics does not describe independent reality but empirical reality, which is a rational picture of reality providing us with reliable statistical predictions concerning atomic events. Reality which we try to describe is outside the reach of human knowledge. We cannot compare our picture with the atomic world which we wish to describe. The reliability of our picture must be judged on the basis of the predictions it gives us.

In this situation it is natural to state that our theory only concerns our knowledge of the (hypothetical) atomic world. This means that the consciousness is unavoidable in our conception of reality.

The innumerable 'new approaches' to quantum mechanics are attempts to avoid this consequence. However, for the creators of this theory it was clear – at least since the 1950s – that *Cartesian dualism cannot be maintained in atomic physics.* This is especially clear when thinking that we can by the choice of the observation method influence the experimental properties of the atomic objects. In certain experimental arrangements we see particle properties while some other methods show that these objects are waves. The free will of the observer influences the experimental properties of atomic objects. Free will is one of the most characteristic aspects of the 'world of spirit.' Thus, we must state that the 'world of spirit' is inseparable from the 'world of matter,' as we experience the latter.

I do not know that anybody else had analyzed this situation more clearly than Pauli. For him, the psychophysical problem was the most important problem of our time. His starting point was physics, the investigation of the 'outer world,' but the strange difficulties encountered in the measurements of atomic physics led him to analyze observations from a psychological point of view, and especially in the light of the depth-psychology of C.G. Jung. The result was the conception of reality I have tried to describe above. It has not been accepted at all by the physics community so far. But now the Pauli-Jung collaboration is beginning to interest people more and more. For physicists in general, it will be very difficult to accept Pauli's thought because it is something fundamentally different from the pragmatic attitude ruling in physics today.

Pauli speaks of physics and psychology as complementary sciences in the same way as particle picture and wave picture are complementary descriptions of the atomic world. Both are needed, and only together can physics and psychology give a realistic picture of the world. I said 'of the world,' not 'of the atomic world' because this picture of reality concerns human knowledge in general, not just quantum mechanics. In fact, the same observational problems as in atomic physics are encountered in all experimental sciences if one begins to think of these questions carefully.

In the conception of reality which Pauli and Jung together tried to develop, the concept of archetype has a basic importance. It crystallizes the idea of the cosmic order which I mentioned earlier and which finds its expressions both in the laws of nature and in the functioning of our psyche. In Jung's psychology, the archetype concept is one of the basic ideas. It has developed over a very long time into an ever more abstract form. Originally it meant the archaic, symbolic pictures which appear in mythology and in the folkloristic tales in different parts of the world in surprisingly similar forms. Jung interpreted them as expressions of the functioning of the unconscious which is based on similar genetic abilities of the human psyche, most probably based on certain inherited structures in our brain and our nervous system. From this original idea of archaic pictures, the idea of archetypes developed into the idea of abstract shaping abilities of the human psyche.

In the idea of archetypes Pauli saw the basis for the cosmic order which he had found in the physical world. According to the old neo-Platonic idea, archetypes are memories in the human soul of the eternal *world of ideas* to which the human soul itself belongs. They are God's thoughts according to which the world is created. Johannes Kepler, who was strongly influenced by the neo-Platonic philosophy, tried to discover the basic archetypes in the cosmic structure of the world. He used the term archetype in a sense similar to Jung, and according to the neo-Pythagorean terminology he also called them the *harmonies of the world*. In his treatise on Kepler's scientific work, which Pauli published in the volume *Naturerklärung und Psyche* (*Jung & Pauli* 1952), Pauli describes the influence of archetypes on the scientific work of Kepler.

But Pauli also tried to study his own unconscious by using the idea of archetypes. He had learned to analyze his dreams according to the Jungian methods, and he continued this investigation of his own unconscious until his death. The Pauli-Jung correspondence (*Pauli & Jung* 1992) contains a description of this activity of Pauli which reveals a new aspect of his personality. Pauli's main aim was to determine some basic archetypes which influenced his thought.

The archetypes which Pauli had in mind were quite different from those in Jung's work. Pauli spoke of the archetypal role of natural integers and of symmetries which are described in mathematical *group theory*. He seems to have continued his scientific work in his dreams in a very interesting and strange way, and the dreams and free fantasies, which he describes in his letters to Jung, open surprising new perspectives. It is not possible to say today what the final value of this little known activity of Pauli will be.

After the Second World War, Jung, like Pauli, emphasized the 'complementary' nature of the archetypes. They belong to the 'world of matter' as well as to the 'world of spirit.' Pauli and Jung saw in this concept the possibility of tackling the psychophysical problem in a new way. They called their conception of reality *unus mundus*: one world. Matter and spirit are just human ways to give a shape to phenomena in an abstract world which is both material and spiritual (mental). Archetypes can be considered the abstract basic elements of this one world. In his final period, Jung emphasized that archetypes are ordering elements of the unconscious, which, simultaneously, must in some way appear in the material functioning of the brain. Pauli began very early to consider archetypes as structural elements in an abstract world which by nature is 'complementary' – both physical and psychic – in a way analogous to the atomic world's being 'complementary,' having a 'both-and' structure which cannot be described by the normal 'either-or' logic. This means that *archetypes are irrational by nature.*

This is understandable because archetypes are ordering factors in the unconscious, which by its very nature is irrational. We cannot describe unconscious processes by strictly rational (mathematical) theories. We cannot define archetypes in the way we are accustomed to doing in science. We can only see some results of their functioning, like the archaic images, which Jung originally defined as archetypes in his psychology. The archetypes can only be described by analogies and similes. They are ordering elements in the depth of the unconscious.

However, Pauli especially emphasized that one can find expressions of archetypes in nature as well. This is the sense in which Kepler spoke of the archetypes, and even Pauli found them in the first place from the physical world. We must remember that both the inner world and the outer world belong to our consciousness, as Fig. 3 illustrates. Everything what we can say about the world is contained in the picture of the world in our consciousness, and it is not surprising that the ordering factors are the same in the inner world and in the outer world. For most people the idea that the inner world and the outer world together form a whole in our consciousness is certainly very difficult to comprehend. We must become familiar, however, with the idea that *the world is one*, and its division into different elements takes place in our consciousness. *We* divide the world into the inner world and the outer world, into subject and object; *we* also divide it into the world of matter and the psychic world. In fact, the world is a whole: both physical and psychical. We must learn to approach the psychophysical problem from this point of view.

This chapter was published (with small changes) in German with the name "Moderne Physik und das psychophysische Problem: Fragen zu einer zeitgemässen Naturphilosophie" in a volume *Philosophia Naturalis* (*T. Arzt et al.* 1996).

14. The Reality Beyond

Can any prophet arise out of Galilee? This people asked two thousand years ago. A similar doubt seems now to be cast on whether physics can have anything to say about the questions concerning man's existence.

In fact, in physics the question which people were forced to ponder in the 1920s, because of the surprising development of atomic physics, has now again become a burning one. The question is, *what is truth?* What kind of knowledge can we believe in if we meet things which make beliefs and ways of thinking problematic although these have been considered as self-evident. Physics has run into the question 'Where is the border line between knowledge and belief?' Many paths lead us now to this border region – although the scientific community obviously is very reluctant to follow them. Physics is not so far from the fundamental problems of existence as people think.

Materialism in the East and in the West

We are living in a time of crisis. What has taken place in the former Soviet Union and in its neighboring states could scarcely have been predicted. "Fortunately something else will usually happen other than what we expect." So once said my teacher of mathematics, Rolf Nevanlinna, at a critical moment of the Winter War – when the breakthrough in Summa had taken place – concerning the uncertain future of our country. This is real wisdom to be applied to the research of the future. Man is incalculable. Irrationality belongs to life.

Some time ago I wondered why the rulers of the Soviet Union insisted so strictly on ideology. I thought that it would be wiser to give up materialistic dogmas and enter into an alliance with the Orthodox Church. I thought that it would only strengthen their system. Probably they were right, however. If the belief fails, the whole system wavers.

In the Soviet Union everything was based on a belief which was founded on Marxism-Leninism. When this belief failed, collapse occurred. The collapse was, of course, due to economics, but the whole system was based on a belief which also gave direction to economic endeavors.

Without belief, action lacks strength. "All life is willing, and will is the same as belief", so formulated Eino Kaila the "Final balance" in *Syvähenkinen elämä* (The Depths of Spiritual Life). (*Kaila* 1943.)

Belief based on materialism was doomed to fail because its basis was an illusion. Only since the 17th century has materialism grown to a really respected ideological view. This is due to Cartesian dualism where matter and spirit are considered as being two independent 'substances' of the world. In this dualistic picture of reality the relation between matter and spirit has become a veil of mist obscuring Western thought. It led to the idea of psychophysical parallelism, the basis of which is untenable. In the natural sciences, spirit – mind – was totally forgotten.

The excellent achievements of the natural sciences strengthened materialism, and the whole idea of spirit (mind) is generally considered unnecessary ballast. In the West, an honest scientist cannot take seriously problems concerning spirit, and this attitude is strictly valid even today. In the Soviet Union, physical sciences were considered as allies of materialism. The fact that they were the basis of technology was considered a proof showing that a reliable conception of reality is to be founded on the investigation of the material world only.

When thinking, in the light of modern atomic research, of the development which has taken place during the last three hundred years, one can see in it a great illusion created by the success of empirical science. One of the basic elements of the Cartesian conception of reality, the *world of spirit* (*res cogitans*), has slowly but surely come to be seen as an erroneous fantasy. In atomic research, however, problems associated with observations take an extremely delicate form, and in this connection the role of consciousness and of free will cannot be neglected – if we do not wish to impose arbitrary (materialistic) preconditions for the conception of reality. Matters which according to Descartes belong to the world of spirit cannot be distinguished from observation processes, and therefore it is necessary to think of the relation between 'matter' and 'spirit' – how spirit influences material phenomena, and vice versa.

We shall come back to this question later. Here it is enough to state that materialism has lost its foundation because of atomic research. Physics has been a deceitful ally for materialism. One must add, however, that physicists in general do not ponder over questions of this nature.

Because of this, the collapse which occurred in the East can be seen as a consequence of the fact that life was based on an illusion. The materialistic ideology gave a wrong direction to the endeavors. The world was imagined to be a machine which can be dominated with the aid of reason. This inspired one-sided values, which at last caused a rebellion of the spirit. The suppressed desire for freedom burst forth irresistibly and tore away the strait jacket used in governing the nations.

But is the Western world then free from materialistic illusion? Far from it. Materialism simply appears here in another form. Free economic competition is the form of life which is recommended in the West instead of the planned economy. 'Market forces' may freely steer the community, including even cultural life. This is full materialism. Material interests and selfishness have risen to incomprehensible esteem.

In the West another form of materialism rules – a form which respects freedom. It has problems of its own, and the depression has reminded us of them. What, in fact, caused this depression and the enormous unemployment in our country? We do not know so far the causes, but a general unscrupulous swindling and selfish misuse of the economic system have certainly played a role.

We do not know yet where the development will lead us now. The value basis of society has been displaced, and it is not possible to foresee what this means. "But seek ye first the kingdom of God, and his righteousness; and all these things shall be added unto you". So we once learned. Now this has been replaced by new truths and new values, and people wish to build a proud new world. Material aims are one-sidedly emphasized, and people forget that man does not live by bread alone.

I do not believe that materialism in its Western form either will endure with the passing of time. *It is not possible to cut off from reality an essential part of it without being led to untenable solutions.* We have arrived at a period when the spirit's existence has to be acknowledged again. The relation between matter and spirit now requires reconsideration. Cartesian dualism has to be replaced by a more satisfactory conception of reality. This I try to illustrate in the following.

A New Conception of Reality

Atomic research, unexpectedly, has found a new point of view for comprehending the relation between spirit and matter. It is the idea of complementary reality – an idea which is used in describing the nature of the atomic world. It is possible to consider matter and spirit (mind) as basic elements of reality from which neither one can be neglected or be reduced to the other one. Instead of the two 'worlds' of Descartes, there is only one world, *unus mundus*, which expresses itself both in the physical and in the psychic phenomena we experience. Let us consider, in more detail, what might be the nature of this one world which appears both as physical and as psychic phenomena and what is the relation between these basic 'substances' of reality. The wise atom has given us a hint. Let us try to understand what this hint means.

Why does atomic physics mean the end of Cartesian dualism and of materialism as well? The vindication of this fact must be repeated again and again because physicists have not made the thing clear for themselves, not to mention philosophers. In the case of physicists, the reason is the pragmatic attitude which helps one to ignore philosophical questions as useless deliberation. On the other hand, philosophers seem not to notice enough empirical results but try to 'analyze them away' by using logical considerations. Natural philosophy which delves into philosophical questions on the basis of natural sciences is today missing in almost all universities and research institutes.

To the founding fathers of quantum mechanics the situation cleared up slowly. In the 1950s they began to use the expression that quantum theory does not describe the atomic world but our knowledge of the atomic world. On a par with the material world there appears another element of reality: knowledge and that which knows, i.e., the consciousness.

In his book *Physik und Philosophie*, Werner Heisenberg writes (according to the English edition):

If one follows the great difficulty which even eminent scientists like Einstein had in understanding and accepting the Copenhagen interpretation of quantum theory, one can trace the roots of this difficulty to the Cartesian partition. This partition has penetrated deeply into human mind during the three centuries following Descartes and it will take a long time for it to be replaced by a really different attitude toward the problem of reality. (*Heisenberg* 1959, p. 75.)

An observation never means just a strictly objective registering of facts but it is a combined result of stimuli coming from the outer world and of the functioning of the human psyche. This functioning of the psyche is, to a considerable degree, unconscious. This creates the 'veil' which hides the reality itself from us. Pauli expressed the same thing by stating that actual observations always contain an irrational element.

These remarks concerning the nature of observations, are not, of course, valid only with respect to atomic phenomena. All observations presuppose a *shaping activity* of the human psyche which for the most part is unconscious. For that reason, human knowledge can only be probable knowledge. Therefore the details of reality remain forever veiled to us. Physics describes the rational features of reality, but the reality revealed by our observations also contains an irrational aspect, and therefore 'reality-itself' remains forever 'behind a veil'.

This is the profound teaching of atomic research. It means that we can describe the world only with human restrictions. *Simultaneously it becomes impossible to make a distinction between the observer and the observed object.* The observer and the observed form during an observation an inseparable whole, the details of which cannot be specified. It is as if the human psyche were mixed in with everything that we can say about an object. There is no 'inanimate nature'. Reality – as far as we can say anything about it – always contains features which one would like to call psychic.

In atomic physics this appears especially in the form of free choices which are characteristic of individual events, when laws are genuinely probabilistic. The view of the world as a machine is an illusion.

Materialism is based on the idea that there is an objective outer world independent of observations. This idea is based only on a belief, not on empirical facts. If reality means something that can be described by using the methods of empirical science, we cannot purge from it the expressions of the human psyche, and this makes materialism very weakly motivated.

Matter – Soul – Spirit

What kind of picture of the one world, unus mundus, do we get on the basis of atomic physics? What is the role of matter and what the role of spirit (Geist) in this new conception of reality?

At first we must agree upon a change in terminology. Cartesian dualism concerns in fact the physical and the psychic elements of reality. It is better to avoid the term 'spirit' or 'mind' in this connection. We shall use, instead the Greek word *psyche* to mean the opposite of matter (or physis), and we say in the following that Cartesian dualism concerns the distinction between physical and psychic.

In their common deliberations over several decades, Wolfgang Pauli and C.G. Jung arrived at the idea of unus mundus which means the conception of psychophysical reality. Pauli came to this idea explicitly from atomic physics, Jung on the basis of his psychiatric activity. Pauli's thought is better known to me, and my exposition will be mostly based on Pauli's articles and letters.

Pauli often emphasized the mystical force which mathematics manifests in guiding the development of physical theories. Mathematics is a creation of the human spirit and, thus, it explains the structure and the functioning of the human psyche. The fact that mathematics helps physicists in such a wonderful way in shaping a picture of the physical world shows that the structure of the physical world, and the functioning of the human psyche are governed by the same cosmic order which is beyond the Cartesian separation of physical and psychic. The idea of a cosmic order is essential in Pauli's thought, and he found that Jung had come to very similar conclusions on the basis of psychiatric phenomena.

Especially after the Second World War, Jung emphasized the physical basis of the unconscious psychic life, and he introduced the term *psychoid* to mean something that is both physical and psychic. This was the nature of Jung's conception of reality. The correspondence between Pauli and Jung in the 1950s was very much concerned with this psychoid conception of reality, and Jung introduced the name unus mundus which had been used, in an analogous sense, by the medieval physician and alchemist Gerhard Dorn.

This one world is so abstract by its nature that we cannot form any concrete picture of it. We can only describe this abstract reality by using similies and analogies. One is apt to call this abstract reality *transcendent*, reality beyond, because its description is beyond the abilities of human reason. We can only have intuitive premonitions of its nature. Physical phenomena and psychical phenomena are some types of projections of this abstract reality, and we can only experience such projections of it.

For Pauli, the model was the atomic world, which for us appears either as waves or as particles, depending on the observational means we are using. It is not possible to form any concrete picture of the atomic world although it is possible to describe mathematically both wave and particle phenomena by using the quantum mechanical formalism.

Thus, Pauli began to speak of a psychophysical world, the physical and the psychic aspect of which are complementary in an analogous sense as the wave aspect and the particle aspect in the atomic world.

Here only some general remarks can be made about this new conception of transcendent reality.

First an important remark concerning terminology: The abstract reality, unus mundus, which we cannot experience directly but only as its physical or psychic projections, I would like to call *spiritual*. Over the course of time the basic reality has often been called spiritual, although this has not been done with reference to the concept of complementarity. According to this terminology, reality itself is spiritual but we cannot directly experience what 'spiritual' means. We have experiences which we call physical and neglect the psychic side associated with them. On the other hand, we have experiences which we, correspondingly, call psychic neglecting their physical aspect. The spiritual source of these experiences remains behind a veil for us.

I have tried to illuminate this with the frontispiece to Part IV although I then try to give a picture of something that cannot be described by pictures. The figure can only be a rough analogy of reality itself.

In that picture, the spiritual reality has been presented as three-dimensional for the simple reason that we cannot imagine more than three dimensions. The physical phenomena have been presented on the x-axis of the plane coordinate system in the figure which, of course, means rough simplification; in fact, we experience physical phenomena in three-dimensional space and in time, which presupposes four dimensions, and in modern theoretical physics often more dimensions, perhaps even infinite-dimensional spaces are needed for the description of the physical phenomena. Here we must use a very crude simplification, however, and everything that is experienced as physical is presented on the x-axis. Analogously, everything that is experienced as psychic has been presented on the y-axis of the same plane coordinate system.

In science as it currently prevails, physical phenomena are strictly separated from psychic phenomena. We can, however, consider phenomena from a more general point of view and say that phenomena always have physical *and* psychic dimensions, and then a phenomenon can be described on the xy-plane of our figure. If the psychic dimensions are of less importance from the point of view of the phenomenon, then we can speak of a physical phenomenon and describe it as a curve for which the y-coordinate varies only a little and the curve is directed mainly in the x-direction. If we wish to leave behind Cartesian dualism, we must, however, describe all phenomena as curves in the xy-plane. We suppose that science will be generalized so that both physical and psychic dimensions will be taken into consideration in all phenomena. The world of science is then the xy-plane of our figure.

In the prevailing physics, this kind of view is found unsuitable. Physicists don't like to abandon the principle that it is forbidden to mix physics with psychology because otherwise the objectivity of physics will be lost. I find this to be a very dogmatic principle. The 'paradoxes' of quantum physics are eliminated in a simple way by considering the role of the consciousness. For creators of quantum mechanics this became more and more clear in the 1950s and they expressed this by saying that quantum physics only describes our knowledge of the atomic world. Thus one has to include the psychic dimensions into the interpretation. (With respect to Bohr, see pp. 41–44, 67–68 above.)

After the deaths of Pauli (1958) and Bohr (1962), the younger generation began to develop interpretations insisting dogmatically that the physical phenomena must be described as purely physical, without mixing psychology into the description. The result is that bookshelves in the libraries are overloaded with literature which describes quantum mechanics in this way, corresponding to the materialistic conception of reality.

Famous is the mystery of the *reduction of state function* in an observation. It has been described on an almost astronomical number of pages and this writing goes on. For those who know this problem we can remark that it is the most simple problem in the world if we state that we get, in the observation, some new knowledge (new information) concerning the system under consideration, and therefore we have to re-estimate our knowlege of the state of the system. This is not acceptable for a materialist because 'knowledge' is here understood as something that presupposes the inclusion of the essentially new psychic dimension.

The World of Spirit

In the frontispiece to Part IV, the plane of science belongs to a three-dimensional space which I have called the world of spirit. I will now proceed to the discussion of this three-dimensional world. It presupposes that we go over from physics to the realm of metaphysics. The aversion to metaphysics among physicists is unnatural because it cuts the roots of physics and makes the opening of really new perspectives impossible. Great revolutions in science always begin from metaphysics.

The third dimension in the figure describes the irrationality of reality. The world of science is rational by nature. The third dimension describes the fact that reality contains aspects that cannot be reached by scientific methods. It is possible to imagine, of course, that there are more than just one irrational dimension but this can only have a symbolic meaning because we are speaking here of irrational matters.

The three-dimensional space corresponds to Kant's *reality-itself*. In physics we we speak of *independent reality*, which exists independent of human observations. As was said above, it contains explicitly an irrational element which leads us outside the world of science. We can imagine that the real change takes place in the three-dimensional 'reality beyond' but in our experiences a change appears as a psychophysical phenomenon. In fact, the present science describes each phenomenon as *either* physical *or* psychic, and one is not allowed to mix these with each other. In every case, the idea of irrational elements is absolutely excluded from the prevailing science – one is not allowed to speak of elements which are not to be reached by scientific methods at all. Such an irrational aspect in phenomena is for instance the effect of free will. It makes life phenomena especially unpredictable. I wish to remind you of the statement of Rolf Nevanlinna mentioned above: "Fortunately, something else will usually happen other than what we expect".

I think that science cannot in the future neglect the irrational dimensions of reality. Scientism which currently governs the thinking of the scientific community with regrettable force flattens down the picture of the world to the *xy*-plane, and exactly this fact creates the illusion about which I wish to warn in this book. It is this very scientism that has in practice been responsible for the strengthening of the materialistic Utopia.

What can we then say about the world of spirit when it lies outside the plane of human experiences? Nothing in the sense of strict science. If one wishes to say anything about it – and I find it important that we try to speak of it – it presupposes a comprehension in a sense which is not restricted by the rules of strictly rational scientific language. It presupposes an intuitive vision and *belief*. Belief associates us with the world of spirit. It gives flashes from the 'reality-itself'. It is regrettable that such metaphysical flashes which have been presented on the basis of atomic physics are usually related to the religions and philosophies of the Far East. Christian theology, so I have understood, has taken a rejectionist stance towards all ideas concerning the world of spirit which are based on science. This has essentially strengthened materialism in science – and continues to strengthen it. Science has become another religion, and I fear the implications which this will have in the course of time.

Perhaps I dare, in spite of the possible criticism of the theologians, to present some hints at how one could try to find solutions to some controversies between Christianity and science which factually exist today. A natural scientist who personally finds Christianity to be the only reliable way to the world of spirit feels these controversies as painful. It is difficult to understand that theologians do not see these questions: They continually emphasize the splitting of reality in the sense of Cartesian dualism, which does not bear scientific criticism any more.

First, it is necessary to remark that the original Christian view seems to have corresponded to the picture of man which is formed on the basis presented here. One has to make distinction between the concepts 'body', 'soul', and 'spirit'. *Spirit is the divine spark in man.* Body and soul are related to the earthly existence of man, spirit is immortal. St. Paul expresses this very clearly in the 15th chapter of his first letter to the Corinthians.

What I have here called the 'world of spirit', is for me simultaneously a picture of God who, finally, is the only reality. I wish immediately to emphasize that this does not mean pantheism because God is transcendent by nature: He cannot be described by using human concepts and He cannot be identified with the universe which science describes. He is the *Reality Beyond*.

What makes it difficult for those who are accustomed to the scientific conception of truth to acknowledge religious truths, is the fact that everything that can be said about the world of spirit is explicitly symbolic. It is to be received more by the heart than by reason. The spiritual essence of man can give to him flashes from the reality which science cannot reach but which can only be reached by faith. As a psychophysical being, man is bound to the world of time and space, but in death this bond is broken and only the immortal spirit remains.

Although God is transcendent, He is simultaneously immanent, ubiquitous in the world. The world of science also belongs to the world of spirit: It is just a section of this – 'three-dimensional' – world. The visible world which we experience via our senses is 'in God'. "For in him we live, and move, and have our being", said St. Paul to the Greeks on the Areopagus. The world is in God and, thus, God is ubiquitous.

Characteristic of this conception of reality is that it is as if nature makes 'free choices' in phenomena between different possibilities. This new freedom contained in changes can be interpreted as God's will. Phenomena are not governed by 'chance' as in the prevailing theories of evolution but by *God's will*. God is present in everything that occurs. He gives direction and meaning to that which takes place.

Supplementary Note:

In this book I have tried in general to follow Pauli's terminology because he represents in the most profound form the original interpretation philosophy of quantum mechanics. In this chapter, however, I have deviated from Pauli's terminology – although not from his conception of reality, as far as I understand. Pauli used the word 'spirit' [Geist] and correspondingly the adjective 'spiritual' in the same sense as I have here used the terms 'psyche' and 'psychic'. Thus, Pauli has used the term 'spirit' in the same sense as it is usually used in Cartesian dualism: One speaks there of the 'world of matter' and of the 'world of spirit (or mind)'.

About the psychophysical, abstract reality, which I have called 'spiritual', Pauli has used the Jungian term *unus mundus*. Earlier he spoke of the 'intermediate world' [Zwischenreich] or sometimes of the 'kingdom of the middle' [das Reich der Mitte], in reference to an old Chinese terminology, or he has used for the 'reality beyond' simply the name X. Essential is that this *proper reality* is transcendent or, as Pauli also often said, the 'world beyond'. Essential also is that it is one. *It is a whole that combines the complementary opposites.*

Jung sometimes, in his correspondence with Pauli, used of this intermediate world the name 'psychic'. He meant that 'psychic' and 'physical' are terms which refer to the transcendent reality and which cannot be distinguished from each other (Jung's letter to Pauli of October 24, 1953). As a physicist, Pauli could not accept this terminology. Unus mundus is, of course, a good name for the transcendent reality but I think that the term 'spiritual reality' or the 'world of spirit' brings us closer to the normal usage of words – and I also think that it is close to the early Christian views.

Both Pauli and Jung emphasize that if the 'soul' is understood to belong to the eternal world of ideas, as in neo-Platonism, this causes a severe confusion in the terminology, especially if one thinks that the world of ideas is good and this property is missing from matter which therefore is evil (*privatio boni* = the lack or absence of good). This has resulted in calling the 'soul' immortal, instead of the 'spirit'.

15. Reality and Values

A stormy century is coming to an end and it is time to stop for a while and think about what important results the 20th century has been able to contribute to the development of Western culture. Since I am a physicist, I am inclined to see my own field as especially important. "Who drives with oxen, he speaks of oxen." So I think that later, when enough time has passed, the importance of some physicists will become more obvious – as was the case with Newton and his achievements in the 17th century.

As long as there is time. I am an incurable optimist, but I cannot deny that I see the future as being overshadowed by severe threatening factors. One can ask, with good reason, whether there will be people in the year 2100 or 2200 who are interested in the cultural achievments of the 20th century. The course of events is wrenching itself from our hands, as in the story of the wizard's apprentice who was able to produce a deluge but was not able to stop the rush of water.

I have tried above to give reasons for the view that modern physics – if the message of the atoms is really understood – can change the basic direction of Western culture in a profound way. The question is, indeed, whether these new perspectives are understood in this world which is becoming more and more superficial and where even truth is harnessed for the exertion of power.

Reality and Absolute Values

Some physicists, and even some philosophers, have presented ideas concerning religion on the basis of modern physics. I shall only mention here the French bestseller *Dieu et la Science* (God and Science), which an old French philosopher Jean Guitton published together with two physicists, Grichka and Igor Bogdanov (*Guitton* 1991). Among people with a humanistic education – including theologians – this has occasioned strong criticism expressed, for instance, in the slogan: "Why should God reveal Himself in a test tube, precisely?"

This kind of remark perverts completely the nature of the matter. Atomic physics uses very effective means in the investigation of the basic structure of matter and presupposes a profound change in our conceptions of both human knowledge and the nature of reality. This change seems to open new perspectives which can eliminate the unfortunate controversy between empirical science and the Christian belief. One should not be afraid of such perspectives or try to belittle them, but should instead try to examine them more closely. Actually, the basic philosophy to which we are led from quantum mechanics is essentially the same as Guitton's and his collaborators' 'metarealism'.

In the 'humanistic language' of the slogan above, I would like to say: "One cannot expect, of course, that God would reveal Himself in a test tube, specifically, but surely one can also find Him when investigating the structure of matter!" We should not forget the general revelation: God speaks also through his Creation. In exact sciences we get the most objective results that a human being can acquire. This is the objective way to God, the subjective way can everybody find in his/her heart. In both cases, God is the same. In the East this is better understood than in the West: Brahman is the same as Atman.

When teaching the natural sciences, one generally describes empirical research as a value-free struggle for an objective truth. Its foundation is composed of the hard facts of empirical research. Natural scientists truly believe that this is the case, and if somebody tries to shake this belief, he soon finds himself spurned by the scientific community.

But is this belief value free and completely objective? During his education, a natural scientist appropriates a certain basic attitude: belief in the omnipotence of the empirical method. In physics, this also includes the Einsteinian belief that reality is rational by its very nature. When atomic research seemed to lead to another view, physicists and philosophers tried to outdo each other in changing quantum mechanics and its interpretation in such a way that the 'epistemological lesson', which Bohr strongly emphasized, would be covered so completely by a complicated logical construction that nobody could understand the nature of that empirical lesson. Most physicists and philosophers believe that reality must be rational.

However, if this precondition is abandoned, quite new views can emerge. The physical and the psychic elements of reality are fused into one inseparable world, and then one cannot avoid questions concerning *values* and *purposes*.

Since the time of Francis Bacon, natural science has endeavored to rule over nature. Research has aimed at the exploitation of nature. This has resulted in the application of sciences in more and more fields and in an enormous rise in living standards – but simultaneously also in increasingly difficult environmental problems and ever more effective means of destruction. Man has utilized his knowledge in order to change natural processes according to his own, selfish aspirations. All the time, the basis of the attitude has been a view derived from Cartesian dualism: that the object of scientific research belongs to the 'world of matter'. We now have good reason to be afraid of a reaction: The one-sidedly misused nature can cause a 'backfiring' which stops this unnatural development where the spiritual reality has been totally ignored.

In his article "Beyond Baconian Quantum Physics," Hans Primas, professor of physical chemistry at the Technical University of Zurich (ETH) points out that so far quantum theory has been applied in the spirit of the classical idea of causality although the structure of the theory would allow quite different kinds of questions (*Primas* 1990). Instead of aiming to predict phenomena, one could direct one's attention backwards in time – one could ask why the world is as it appears to us today:

Backward deterministic and forward purely nondeterministic processes are appropriate for the description of mechanistic processes of Baconian science while forward deterministic and backward purely nondeterministic processes correspond to finalistic processes. In the bio-philosophical literature it is often admitted that teleological thinking may be useful but that any teleological process always can be transformed into a mechanistic one. In the framework of the proposed precise definitions such simple-minded equivalence is certainly not possible. A genuine teleological explanation may be much simpler than a cybernetical backward deterministic explanation using a much enlarged universe of discourse.

It is well-founded to think that natural science could ask questions of a quite different kind: ones that do not aim at the exploitation of nature but at the understanding of nature. While contemporary science puts disproportional emphasis on the question of '*how* phenomena take place', one could put a greater weight on the question '*why*'. It is true that then one perhaps cannot presuppose that everything can be described by using mathematical language and strict logic. One needs science which also accepts metaphysical views and intuition. It is necessary to pay attention to the subject – the scientist himself – and to the process of shaping (gestalting) in the generation of knowledge. If we are searching for the purpose and aim of natural processes, we cannot avoid questions which concern the foundation of values.

How we estimate things depends on our basic views of the nature of reality. The present science one-sidedly emphasizes reason, and the result has been a materialistic conception of reality while spirit has been banished from it, to some kind of 'other reality'. We need a new kind of natural science which acknowledges the reality of man's soul and the ubiquitous existence of a spiritual element in the world. Then we will also put things in science into a new ranking order – we will be asking other kinds of questions.

On my bookshelf I have a volume *Tiede ja etiikka* (Science and Ethics) produced with the support of the Academy of Finland. Its discussions seem rather unfamiliar to me. The problems which are discussed are important; the development of science has created them and continuously creates similar questions. I just feel that something very important is missing from the book. If ethics means merely intellectual analysis, the very foundation is missing from it. It is not possible to separate the ethics of research from the general principles concerning human morals. This is the actual problem of our time. When the role of religion has weakened to such an extent as is the case in the West, the result is a crisis of values. Intellectual evaluation is not sufficient if it does not spring forth from an inner conviction which forces one to even risk ones life for the values which one feels to be inalienably important.

Science is not a sufficient basis for values. It does not reach the absolute values hidden in the transcendent reality which is deeper than the reality reached by science. The foundation of morals is not in knowledge but in belief. The materialistic concept of reality which does not acknowledge the existence of the human soul and of the spiritual element in reality does not offer any durable basis for morals.

The conception of reality prevailing in science today leads to the denial of absolute values. Any search for them is condemned and held to be dangerous, and religions are accused of deep rifts among their adherents. Although these rifts cannot be denied, is not the very reason for them human selfishness and lust for power? Religion is sometimes only used as a means to achieve some other purposes. However, all great religions emphasize the love of ones neighbor and the esteem of life. Religions become dangerous only if they put more weight on orthodoxy than on the basic human values which form the heart of their teachings.

The emphasis on orthodoxy springs from human self-assertion, and it leads to controversies between religions and religious groups. Therefore, religious fundamentalism is a terrible phenomenon, and it is necessary to fight against it. The actual struggle against religion which has appeared in the Western countries and even more in the officially atheistic countries in the East, has been based on superficial views and, in particular, on the error of materialism: The spiritual aspect of reality is considered to be of secondary importance. *An erroneous conception of reality distorts the basis of values.*

Science and Values

But how does one find the 'absolute values' on which morals, in the end, have to be based? This is the question of questions, indeed. They cannot be found by the methods of present science; in fact, the overemphasis on intellectual requirements can preclude the possibility of finding them. They can only be found in the transcendent reality which cannot be reached by science.

I don't know any better rule for finding the absolute values than an honest search for the truth which is deeper than the scientific truth. A good starting point is this: "Thou shalt love God over all and thy neighbor as thyself." This is a superhuman ideal, but it gives the direction. It does not mean any orthodoxy or condemning others because of a wrong belief; it is based on love.

Thus, I have arrived at the teachings of Christianity. I do not do it by regarding lightly the wisdom of the Eastern religions or in general underestimating the support which other religions can give when we try to get in touch with the transcendent reality which is the basis of existence. Personally, I just don't see that any religion besides Christianity could become so familiar to me that it could help me in truly difficult decisions. Without religion, society, like the individual, is adrift in the world. The secularization of Western societies is a dangerous matter, and it has caused the disintegration of the basis of values and the enormous increase in selfishness which threatens to destroy the foundation of this form of society. In addition, highly developed destruction methods endanger the whole existence of the human species. The exaggerated confidence in science has significantly supported development in this direction. Therefore, it is important to make it clear that science does not reach the whole of reality – that the basis of existence is in the transcendent reality which can be reached only by faith.

Does this mean that we have to make a compromise with the rationality of science and try to construct a new science which is partly rational, partly irrational? People usually understand the remarks about the new conception of reality in this way, and the introduction of irrationality into science is firmly rejected. However, the irrationality of reality does not presuppose that we introduce irrationality into science.

Quantum mechanics is an excellent example of a rational theory and of the strength of the rational methods. However, in its *interpretation* we meet the irrationality of reality. This presupposes a change in the conception of causality. Deterministic causality has to be replaced by statistical causality. The quantum mechanical description does not reach reality itself but gives only knowledge of the rationally describable aspects of reality – it gives only statistical predictions concerning observable phenomena. What this will mean with respect to physics has not yet been sufficiently clarified.

In any case, we cannot belittle the use of mathematics in the description of physical reality. It is precisely mathematics which has guided physics to ever new achievements – for example, to the creation of quantum mechanics. Galileo was right when he said that the book of nature is written in mathematical characters. However, in the 17th century scientists began to believe that it was possible to govern nature *completely* in this way. Nature has now clearly pointed out that it is not a machine. There is also an aspect of irrational freedom in nature which cannot be governed by the methods of rational science. This freedom which appears in the 'random' scattering around the expectation value predicted by the statistical law is an expression of *creativity*. This fact must now be acknowledged in science, although it means certain limitations for science. Science should not try to answer questions which belong to the realm of faith.

Scientism does not acknowledge these limits. It has created scientific fundamentalism. Religious fundamentalism prevents people from seeing clear empirical facts which concern age determinations and the evolution process in organic nature over billions of years. Scientific fundamentalism, instead, denies the belief in creation and in the possibility of teleology inherent in the evolutionary steps – which is the basis of the unbelievable purposefulness of organic nature. Scientific fundamentalism accepts only the 'religion of chance' which denies the meaningfulness of questions which could help us to comprehend the nature of life as we meet it. Scientism aims to govern nature and is not interested in the choices which are characteristic of life and creation. These questions cannot be explained by scientific methods. Besides science, we need faith which helps us to form a harmonic picture of man, reality, and of man's relation to reality.

If we, on the basis of the scientific picture of the world, wish to understand questions concerning existence and its purpose, we cannot consider 'chance' to be the only scientifically acceptable form of choice. We have to ask the question 'why' – for example, why have choices in the course of time occurred as they have? We should try to find features in the world which help us to understand why the result of evolution is as it is. The answers which contemporary science gives concerning evolution are based on the erroneous materialistic conception of reality; they don't describe the evolution of the real world.

One has to approach these questions without the limitations of the mathematical sciences. The struggle for objectivity has resulted in not regarding the problems of the subject himself with sufficient concern. This perverts the conception of reality and directs the interest in a one-sided way. Life cannot be understood if man does not try to understand himself and his relation to reality.

And here a scientist – if he is honest – necessarily encounters God. God is the basis of existence. When searching for reality we meet Him, and this is the way to the world of absolute values.

Personal God

In describing the basic importance of observations for the scientific picture of the world, the well-known American physicist J.A. Wheeler, an excellant expert of the Copenhagen interpretation, illustrates the nature of the physical reality with a dialogue between the universe and a human being (*Wheeler* 1980, p. 37):

Universe: I am a giant machine. I supply the space and time for your existence. There was no before before I came into existence, and there will be no after after I cease to exist. You are an unimportant bit of matter located in an unimportant galaxy.

Human being: Yes, oh universe, without you I would not have been able to come into being. Yet you, great system, are made of phenomena; and every phenomenon rests on an act of observation. You could never even exist without elementary acts of registration such as mine.

Here Wheeler calls the physical reality the universe. It is the object of a physicist's belief. A physicist's whole endeavor and work is based on the belief that there is something that expresses itself in our observations. This belief is the foundation and basis of empirical research. Without this belief, there would not be any natural science.

But what if we wish to dig still deeper: if we try to say something about the transcendent reality which appears to us both as physical phenomena and as inner experiences? *It* is the origin of existence and of our inner experiences – that which created the universe and the human being who is able to create a picture of the universe and of himself. This transcendent reality reveals itself both in the structure and phenomena of the outer world and in the depth of the heart of the tiny human being. A scientist is searching for this reality when he honestly searches for truth. The same reality is also experienced by every person, when he awakens to search for the basis and purpose of his own existence – and the values to make distiction between right and wrong.

I call this innermost reality God. I experience Him as a person, and I am in dialogue with Him when I search for truth using the methods of science. Because science is dialogue. One must put a clear question to nature; then it gives a clear answer. What kind of questions we ask depends on our view of the nature of reality. A scientist always has some ideas which give a form to his belief in independent reality. The more unprejudiced the thinker is able to be in his work, the better are his chances of finding something important.

Science is one way to approach God. But it can also lead away from God – if the scientist associates his own expectations with his belief: if he trusts too strongly in his own imagination concerning reality. Western thought has been misled by too strong a trust in the rationality of reality.

Science is not the only way. Man can also find God in his own heart. Religion helps man in this search. But even religion always contains secondary visions. One should not stumble on them. One has to search for a personal God who speaks to man personally. Then He can become the foundation of life which also helps to put values in the proper order when it is difficult to find the right way.

Because *God is and He is present* – everywhere and in everything.

16. Hubris and Punishment

A Personal Vision

This is a book about the search for reality. I have written it as a natural scientist, trying to point out that atomic research forces us to re-evaluate many basic questions concerning human knowledge and existence. I could not avoid religious matters because the basis of Cartesian dualism was found to be untenable in the light of the criticism rising from the psychic aspects of observations which has been necessary in atomic research – although people have not generally understood the situation. The question is whether we are ready to dig so deep that even the problems concerning the borderline between knowledge and belief come into the limelight. The critical analysis of these questions will necessarily influence the basis of Western culture.

So far, natural sciences have been based on the belief in the existence of an objective reality which is independent of our perceptions and in its describability by the rational methods of science. This belief is so categorical today that it is not possible to even discuss its reliability. However, we have seen that this belief is not compatible with the Copenhagen philosophy concerning atomic theory.

Number mysticism was a strange field for me until Pauli and Jung aroused my interest in it. I have begun to understand that, besides the quantitative properties which are considered in arithmetic and in number theory, numbers also have qualitative properties. These qualitative properties reflect the very origin of the number concept and simultaneously the unconscious shaping activity which, in the end, is the basis of our knowledge of the world.

Number mysticism has also opened to me the doctrine of the Trinity, which had long remained a closed book to me.

Now I find it necessary to add to these considerations something where I cannot avoid a more personal emphasis. At the beginning of the book, I said that I am in Purgatory. Much has changed and is changing – both in me and in the world. Perhaps the views which begin to take shape in my mind and the basis of this process can have more general importance. Anyhow, it is soothing to tell my worries to others, to point out to others the dangers which burden me. Therefore, finally, I give this personal vision of where we are and where we are going.

Wolfgang Pauli has in a decisive way influenced the course of my thinking. Let us first listen to him.

Science in Western Culture

In his 1955 Mainz lecture, "Die Wissenschaft und das abendländische Denken" [Science and Western Thought], Pauli describes the role of science in Western culture as follows (translated by Robert Schlapp, in *Pauli* 1994, essay No. 16):

Western thought as a whole has always been influenced by the near and far Asiatic East. It seems to be agreed, however, that science, more than anything else, is really characteristic of western civilisation. Mathematics and natural science are specially distinguished from man's other intellectual activities by being teachable and verifiable. Both qualities demand a lengthy and in part critical elucidation. By teachability I mean communicability to others of trains of thought and of results, made possible by a progressive tradition, in which the learning of what is already known requires an intellectual effort of quite a different kind from that required for the discovery of something new. In the latter process the creative irrational element finds more essential expression than in the former. In science there is no general rule for passing from the empirical material to new concepts and theories capable of mathematical formulation. On one hand empirical results provide stimulus for trains of thought; on the other hand thoughts and ideas are themselves phenomena, which often arise spontaneously, to undergo subsequent modification when brought face to face with the observational data. It is not always possible to check by experiment every separate assertion of a scientific theory, although the system of thought as a whole must, if it deserves the name of a scientific theory, contain possibilities of a check by empirical methods. This is what constitutes its verifiability.

Teachability is common to exact science and to logically provable mathematics. The possibility of mathematical proof, and the possibility of applying mathematics to nature, are fundamental experiences of humanity, which first arose in antiquity. These experiences were at once regarded as enigmatical, superhuman and divine, and contact was made with the religious atmosphere.

It is here that we meet the fundamental *problem of the relation between knowledge of salvation and scientific knowledge.* Periods of dispassionate research on critical lines are often succeeded by others in which the aim is to try to include science in a more comprehensive spiritualism involving mystical elements. In contrast to science, the mystical attitude is not characteristic of the occident (Abendland); in spite of differences in detail it is common to occident and orient. In this connection I may at this point refer the reader to an excellent book by *R. Otto*, "West-östliche Mystik" (Gotha, 1926) which makes a comparison between the mysticism of *Meister Eckhart* (1250–1327) and that of the Indian *Shankara* (about 800), the founder of the Vedanta philosophy [*Otto* 1926]. Mysticism seeks the unity of all external things and the unity of the inner man with them; this it does by seeking to see through the multiplicity of things as illusory and unreal. Thus there comes about, stage by stage, man's unity with the Godhead – Tao in China, Samadhi in India or Nirvana in Buddhism. The last-named states are likely to be equivalent from the western point of view to the ego-consciousness. Thorough-going mysticism does not ask "why?" It asks "How can man escape the evil, the suffering, of this terrible, menacing universe? How can it be recognised as appearance, how can the ultimate reality, the Brahman, the One, the

Brahman, the One, the Godhead (no longer personal for *Eckhart*) be seen?" It is however in keeping with the spirit of western science – in a certain sense one might say with the Greek spirit – to ask, for instance, "Why is the One mirrored in the Many? What is it that mirrors, and what is mirrored? Why has the One not remained alone? What originates the so-called illusion?" In his book, mentioned above, *Otto* pertinently speaks (on p. 126) of the "concern with salvation, which starting from certain situations of calamity, found given beforehand, seeks to alleviate them, not however to solve theoretically the problem of whence they come; and which is content to let insoluble problems lie, or cobbles them up as best it can with scanty auxiliary theories". I believe that is the destiny of the occident continually to keep bringing into connection with each other these two fundamental attitudes, on the one hand the rational–critical, which seeks to understand, and on the other the mystical–irrational, which looks for the redeeming experience of oneness. *Both* attitudes will always reside in the human soul, and each will always carry the other already within itself as the germ of its contrary. Thus there arises a sort of dialectical process, of which we do not know whither it is leading us. I believe that as occidentals (Abendländer) we have to commit ourselves to this process, and recognise the pair of opposites as complementary. We cannot and will not completely sacrifice the ego-consciousness which observes the universe, but we can also accept intellectually the experience of oneness as a kind of limiting case or ideal limiting conception. While allowing the tension of opposites to remain, we must also recognise that on any path to knowledge or salvation we are dependent on factors beyond our control, which religious language has always designated as Grace.

The discrepancy between the ways of knowledge and salvation which Pauli describes here became enormously intensified in the 17th century. Natural science became fully conscious of its strength, and, as Pauli said, "they went a little too far in the 17th century." The world was imagined as a giant machine governed by deterministic laws, and the 'world of matter' was considered as being completely describable by the methods of science. Inspired by this belief, natural science has proceeded from one victory to another. There are many who believe that no other way is needed. Science has become a religion, and the obligation of scientists is thought to be to uncover the dimness and the errors of the traditional religions.

In this century the Western world has inceasingly judged science to be more important than the way of salvation. It is believed that all matters can be settled with the aid of reason and science. One imagines that man begins to govern the factors which Pauli mentions at the end of the passage quoted above, and speaking of *grace* is found to be quite unnecessary.

However, the application of science to the domination of nature is beginning to have dangerous side-effects. Rationalists are also worried, for good reason, about the unexpected strengthening of mysticism in the heart of the developed Western countries. Something goes wrong, in an inconceivable way. Even in the midst of intelligent Europe, evil-boding crises have broken out; controversies which had already been forgotten have surprisingly become aggravated. People ask what all of this means. And in the background, one should remember the risk of nuclear war. This was almost thought to have subsided, but if the conflicts escalate, they can break loose from our hands.

Personally, I see behind all of this a religious crisis, which follows from the one-sided direction of Western culture. One has cut off something very important from the conception of reality. Western countries are in the grip of a hubris of rationalism. It is thought that reason can replace faith. Some remarks may illuminate how I judge the situation.

The Power of Scientism

In modern times, the deterministic conception of causality has generally resulted in an attitude which I have above called scientism. The neo-Platonic belief that God has created a *perfect* world led in the 17th century to a scientific picture of the world in which there was no place for God. This is illustrated by Leibniz' reaction when Newton wrote that natural laws cannot be absolute and that God can, when necessary, interfere with the course of matters in the world. As Leibniz writes in his first letter to Samuel Clarke (November 1715):

Mr. Newton and his followers have also an extremely odd opinion of the work of God. According to them God has to wind up his watch from time to time. Otherwise it would cease to go. He lacked sufficient foresight to make it a perpetual motion.This machine of God's is even, on their view, so imperfect that he is obliged to clean it from time to time by an extraordinary concourse, and even to mend it, as a clockmaker might his handiwork; who will be the less skilful a workman, the more often is he obliged to mend and set right his work. According to my view, the same force and vigour goes on existing in the world always, and simply passes from one matter to another, according to the laws of nature and to the beautiful pre-established order. And I hold that, when God performs miracles, it is not to satisfy the needs of nature, but those of grace. To think otherwise would be to have a very low opinion of the wisdom and power of God.

The idea of determinism precludes supernatural things from the scientific picture of the world, and this has resulted in the confidence that all problems concerning reality can be solved by the rational methods of science.

Atomic research has signified the end of determinism. An indeterministic aspect has appeared in nature, and this changes the foundation of the scientific picture of the world. Physicists are almost unanimous about this aspect. If this is the case in a science where causality has been most accurately verified, then one has reason to be critical with respect to the possibility of determinism in other empirical sciences as well.

Indeterminism, however, changes the scientific world view in such a profound way that its implications are confronted downright passionately. Not even the physicists who have fully accepted the Copenhagen interpretation seem to be ready to give up all patterns of thought based on determinism. Statistical causality, which replaces deterministic causality, implies the irrationality of reality, but this is not yet accepted in physics, and even less in the other fields where people have not even become convinced that it is truly necessary to abandon determinism. This matter concerns *scientism*, which does not acknowledge any limitations of principle to scientific methods.

Therefore, people develop cosmological theories which presuppose that the development of the universe, even its beginning – in religious language, its creation – can be described in the sense of traditional realism and using mathematical theories. In biology people refuse to acknowledge teleological arguments in the description of the evolution of life, and evolution is used as an argument against the belief in creation.

It is surprising how categorically the habits of thought based on determinism are defended. Logical argumentation does not seem to help. If somebody departs from the habits of thought characteristic of scientism, this is   immediately interpreted as deceiving science.

Cartesian dualism has created such deep controversies that these must lead to severe conflicts, sooner or later. Scientists are not at all tolerant when their unconscious basic beliefs are at stake, and today this includes an unlimited belief in the ability of science to solve all 'real' discrepancies and problems. This belief guides the development of science into questions which in the light of our discussion are caused by an antiquated conception of reality. Such theories can be compared with the *epicycle theories* which were eagerly developed before the Copernican revolution and which faded away when the new conception of the universe was adopted. However, the roles have changed as compared with the 17th century: now scientific institutions are defenders of the traditional belief.

Science has a dogmatism of its own, and the referee system developed for its defense eliminates from the scope of scientific discussion those who try to criticize its basic dogmas.

Selfishness

The materialistic view of the world leads to an emphasis on selfishness in social life. Spiritual values are eclipsed by material goals. In this atmosphere man feels unwell. So a feeling of dissatisfaction is strengthening, and in the East this has caused the whole social system to collaps. I fear that something similar will take place in the West: upheavals caused by the faltering of the basic faith which supports the society and its value system.

The massive unemployment is a hint of the difficulties to be encountered in the future. I am acquainted with these matters only from the Finnish viewpoint but it is hardly different from what has happened and will happen elsewhere. Characteristic of capitalism is the fight for money and power, and in an ever more industrialized society this seems necessarily to lead to the people becoming divided into those who succeed and those who are pushed aside. I am not trying to present any social theories. I just wish to express my strong dislike of this development, which is a result of the ruling social attitudes and which so strongly undermines the human value of many people. It is not acceptable that people who have honestly served society can abruptly lose the results of their endeavors and be pushed aside for reasons which are caused by an attitude that rewards selfishness and recklessness. This leads to conflicts in which the basis of social life crumbles. These conflicts are already very strong in all fields. We are beginning to accept them as unavoidable.

But there are also other moral problems. Because of materialistic values, the status of the family has become weaker. Sexuality, which should be a stabilizing factor in society, takes forms which disturb its balance.

Although feminism was needed to reform the status of women in society, it has also caused changes which have perverted love into selfishness. What should be the cornerstone of a happy life has become a means of pushing selfish interests forward. *Love is giving not possessing.* Quite something else is the message of the media day after day. "Sex is marvelous." It is a means of enjoyment – and when it is taken in this sense, the result is an increase in divorce, the shattering of children's lives, and many kinds of outrages of human values.

Christianity is often accused of an unnatural prohibition of sexuality. Maybe we are now entering an age when sex is free and people can fully enjoy their life. So people imagine. The result is an increasing number of unhappy people, especially young people, who don't find a safe basis for their life. When we meet sexuality at every step, when it is even brought into living rooms in the TV news, then the physical side of sexuality is emphasized at the expense of its spiritual side and, simultaneously, destructive forces are stirred up.

Personally, I must say that without the backbone which Christian morality gives us, I would have gone astray in this atmosphere. Now that life begins to be behind me, I can only state that I am happy to have had this road sign at my disposal.

Sexuality is an invaluable power source, which is also an essential basis for cultural evolution. However, the creative spiritual activity can use this source only if its bursts are under control. "Fire is a good hired man but not a good boss." The free sexuality of today lavishes this power source on secondary goals. This results in superficiality which creates something resembling science or art, but the truly great creations can come into existence only at the cost of renouncement and self-control.

The 'epicycle theories' of modern physics, which I mentioned above, are an example of the nature of the science which is characteristic of our time. Physics produces abundant results, it is true, but the truly great ideas were found at the beginning of the century. Since then, physics has been stagnant in the great basic problems. It is typical that the ideas presented by Wolfgang Pauli, which could open quite new perspectives, do not interest anyone and, indeed, they are fanatically repressed, because they would presuppose a profound change in the basic attitude of physicists. The contact with reality is weakened. Most important is that one 'gets results'.

It is as if a similar disease plagued data processing, for instance. It continuously creates new formal problems without sufficient contact with reality for which methods are developed. The general fancy seems to be that data processing machines can 'almost totally' replace the human psyche sooner or later. The idea of such abilities, characteristic of human beings, as *intuition* and *conscience* has grown dim. The spiritual dimensions of reality are disappearing from the mental horizon.

One cannot deny, of course, that permanently valuable science and art are also being created in our time. However, the feeling of superficiality is very strong. One thing I would especially still like to emphasize: Sibelius' music would not have been created without Aino Sibelius. She offered her life to make it possible for Jean Sibelius to concentrate totally on his creative activity. Creative work does not thrive in an atmosphere of selfishness; it thrives on love, which is giving, not possessing.

The Empty Tomb

At Easter 1994, a discussion was raised in Finland concerning the historical reliability of Christ's Resurrection. As the leading newspaper of the country, Helsingin Sanomat, wrote on the morning of Good Friday (1.4.1994) in its lead editorial:

Many conservative theologians fear that the whole basis of Christianity will falter if one tries to shake this doctrine. ... Recent investigations seem, however, to prove that stories about the empty tomb and about the revelations of Jesus in Jerusalem and in Galilee get more and more concrete details the later the text in question is. The statements timed closest to the very occurrences point, at the most, to apparitions experienced by Jesus' followers. ... The conclusion seems to be that the story of the bodily Resurrection of Jesus is – like many stories of the Gospels – generated only afterwards on the basis of legends, in order to strengthen the faith.

The newspaper bases its article primarily on the books by the German professor of theology Gerd Lüdemann and, especially by the Finnish docent of theology Matti Myllykoski (*Myllykoski* 1994).

It is good that a discussion of these matters is created, because there seems to be a very musty atmosphere regarding such questions within the Church. An honest seeker of truth very easily gets the mark of heretic. But is this not an expression of the prevalence of a very great uncertainty about the nature of religious truths? I know that I am interfering here in a very delicate theological discussion, but I feel a need to present my view as a representative of science, who in his own search for truth has met similar problems concerning truth which here, according to some views, threaten the basis of the Christian faith.

I must plainly state that the editorial of Helsingin Sanomat on Good Friday morning hurt my religious feelings, even though religious orthodoxy is repulsive prevalent in our country – and very generally elsewhere, too – and which declares all search for truth that does not follow the forms and methods of scientism to be 'unscientific'. It is precisely because of this attitude that the idea of the 'irrationality of reality', which is essential for Pauli's conception of reality, is condemned as unscientific: This conception presupposes that science has limitations which a rationalist is not willing to acknowledge.

The 'scientific fundamentalism' of rationalists corresponds exactly to religious fundamentalism. Such sharp attitudes have created the unnatural controversy between *evolutionism* and *creationism* in America. The result has been that the scientific world has entrenched itself within a conception of science stipulated by a rationalistic dogmatism, which also has a confining effect on the development of science itself. Simultaneously, the scientific community becomes increasingly estranged from religion: From the basis of scientific education, religion is thought to be unfair.

Much confusion is caused by applying to religion criteria of truth used in rational science. Even the representatives of religion themselves often seem to do so. If we try to establish contact with a reality which is transcendental in the sense that human reason is not able to describe it, then truth is not merely a question of reason. It is not important whether Jesus was buried in a common grave or in a private sepulchre owned by Joseph of Arimathæ'a, but if one raises doubts about the reality of the Resurrection of Jesus Christ, then this concerns the basis of the Christian faith: One wishes to say that rational arguments show that this is not possible and, thus, an educated person cannot believe such stories.

How could there, among the clergy elite of Jerusalem and the immensely rich landowners, on the whole, be a supporter of a Galilean prophet? Even if this possibility were accepted at the level of principle, it would be difficult to imagine how he could be in contact with Jesus' activity and become convinced of his message. It is also strange that he can afford to give a sepulchre hewn in a rock as the last resting place of a Galilean prophet who was found politically dangerous and sentenced for having sought the status of the Messiah. ... The general practice was that people sentenced as criminals were buried in a common grave reserved for the executed. If the problems concerning burial stories mentioned above are taken seriously, it is difficult to come to any other conclusion than that this also took place in Jesus' case. (From *Myllykoski* 1994.)

These kinds of details do not decide whether we believe in the Resurrection of Jesus – and they are scarcely written in that sense. It is a question of a *mystery*. Reality, which is not reachable by human reason, has come to the world of space and time in the shape of the Son of Man, in order to tell people something about himself in a manner they can comprehend. God has taken a man's shape. The beginning and the end of Jesus' life, in particular, are mysteries where something that cannot be described by any human language takes place. The evangelists have explained these mysteries as well as they have been able to do so. Certainly it is an imperfect description of events which surpass the abilities of human description.

imperfect description of events which surpass the abilities of human description. Something has taken place that according to the human experience can never occur. If because of this one concludes that the story cannot be true, then the Christian faith is replaced by scientism: Only that is acknowledged as true which can be proven by the methods of rational science. The belief in the Resurrection, however, is not a matter of reason but a matter of heart. There are matters which can be experienced as true although they cannot be proven by scientific methods.

The Birth, the Resurrection, and the Ascension of Christ are matters which surpass the limits of the human experience. As truths of faith, they cannot be shaken by using rational arguments.

Christ has truly been resurrected! This truth has been sustained and will be sustained over millenia.

What is the Truth?

Atomic research has caused difficulties for the traditional realism. Reality has disappeared – perhaps into infinity. Human thought cannot reach it. What, then, can we call a truth? The basis of the scientific conception of truth is not clear any more, since we do not see any reality to which our theories should be compared.

Referring to the frontispiece to Part IV, I would like to say that the question is whether we can know anything of the reality-itself – or of 'spiritual reality', as I have called the three-dimensional reality of that picture. A rationalist presupposes that we must limit our talk about reality to matters which can be verified by using scientific methods, i.e., using the rational language of the xy-plane in the picture. The vehement opposition of positivists to metaphysics was in its time a typical expression of this attitude, and the same attitude is still very strong within the scientific community, imposing limits on scientific discussion – and on the support of research.

But what if one then cuts out essential aspects from reality? This is indeed the case, I think. A normal person who has not enjoyed a proper scientific education feels very clearly that science does not reach the whole of reality. From science he vainly seeks an answer to his deepest questions and the values he needs in his most important decisions. The truths of science do not suffice for man. Reality cannot be truncated according to rationalists' wishes.

Today people need some knowledge of the reality-itself which in our picture was three-dimensional. The 'plane of science' alone is not sufficient. We have to take into account the irrational dimensions of reality. For this, belief is necessary. In fact, even science presupposes certain beliefs which usually remain unconscious. Seldom can a scientist analyze as clearly as Einstein the beliefs which form the basis of his work. However, it is these beliefs that determine the direction of the endeavor.

Even in science one needs the insights that flashes of intuition bring to our consciousness – sometimes at a moment of deep contemplation. In religion and in the arts they are still more obvious. The truths of religion and science are not as different by nature as is often claimed. In both cases the question is of the *comprehension of reality*. The deeper we wish to comprehend the very essence of reality, the more clearly we are in the realm of religion. Purely rational knowledge, based only on objective facts, is an illusion. In fact, the 'world of science' in the frontispiece is not strictly two-dimensional, but even it always has some 'irrational thickness'. How much and what kind of 'irrationality' is allowed in science remains a more or less unconscious matter. This is a question concerning the relation between science and religion.

Without religion, culture does not have vitality. Even science cannot live without metaphysics, and metaphysics is a bridge between science and religion. It gives the main direction to scientific work, and unconsciously this is connected with religion supporting culture. Characteristic in this respect is the conception of time. The belief in creation has produced the linear time of modern science.

The question now is whether Christianity will continue to be the basis of the search for truth in the West. I believe so – and I don't see any other possibility. However, then the Christian theology must be capable of a renewal, which is needed today. This does not mean any 'modernization' of Christianity but a search for its own roots – in a dialogue.

Last Judgment

Upon hearing news from the former Yugoslavia of cruelties which should no longer be possible in civilized Europe, one must wonder at how the protection that culture gives us is still so weak. When a possibility arises, the animal instincts of man burst out unhindered and destructive. Whatever might happen if conflicts break loose on a greater scale.

For me it has been difficult to understand talk about the last judgment and hell. This is not compatible with an image of God where His basic attribute is love. It is true that again we are speaking of matters which cannot be described by human language. The last judgment has to be understood as a metaphor, an image to give us some idea of what will happen before God creates new heavens and a new earth. But even then: Would God sentence the major part of the people to eternal hell?

The reality of war has forced me to think that perhaps the talk of the last judgment is true, nevertheless. Free will – the choice between heaven and hell – has been given to man. What did the Lombardian Marcus say to Dante on the mountains of Purgatory?

... Ye, who live,
Do so each cause refer to Heaven above,
E'en as its motion, of necessity,
Drew with it all that moves. If this were so,

There should be joy for virtue, woe for ill.

... Thus the cause
Is not corrupted nature in yourselves,
But ill-conducting, that hath turn'd the world
To evil. ...

Man himself creates hell on earth. Now he has sufficient means for that. Man has the choice: There is a road to the truth and a road that leads to damnation.

In the grip of the hubris of reason Western countries seem to go astray. It is difficult to change this course because the desire for material good dominates and reason constructs nice illusions of reality without noticing that the reins have been taken by the devil himself.

Since God is the God of love, I interpret the last judgment as an event where the error is burnt away and the Truth alone remains. But of this event the Book of Revelation gives a terrible description.

Hubris is always followed by punishment. Therefore we have reason to fear the future.

The Gospel of Love

God is, and He is Truth and Love. This is the heart of the Christian belief in God. However, the created also includes that which distinguishes it from God – the devil. There is the tree of the good and bad knowledge. Freedom belongs to creation: Without freedom, creation is not possible – but freedom also opens up the possibility of error.

Error and injustice disappear when everything that distinguishes the creation from God is burnt away and only truth and love remain – God who is the origin and the root of existence. "And now abideth faith, hope, charity, these three; but the greatest of these is charity."

Appendix
Italian Guests

Learned discussions

The two principal works of Galileo report on the previous discussions which the interlocutors – Sagredo, Salviati, and Simplicio – had four hundred years ago about the reliability of empirical methods. The picture above is from the title page of Galileo's second principal work, presenting three other famous scientists who also discussed the nature of reality: Aristotle, Ptolemy, and Copernicus. (By courtesy of the Helsinki University Library.)

Reality

In the Finnish edition of this book, Part II has the title "Italian Guests" and describes the difficulties which the philosophy of quantum mechanics encountered in Finland in the 1980s. In fact, the situation has not changed very much. This appendix is the translation of the first chapter of this original Part II.

Three famous Italian interlocutors, Sagredo, Salviati, and Simplicio, recently held in secret an interesting symposium in Helsinki, in a room of the restaurant Vanhan Kellari. The meeting was organized by Mr. Spiritualist, and we are publishing here his report of the discussions which he recorded with the aid of modern apparatus in December 1985. [The report was earlier published in Finnish in *Laurikainen* 1987.] The meeting was opened by Sagredo.

Sagredo: Dear friends! We have paid attention to the discussion concerning the foundations of natural philosophy which has been taking place in Finland. Since this discussion has certain analogies with the corresponding notes which our famous Academician Galileo Galilei has described, we accepted with pleasure the invitation of Mr. Spiritualist to deliberate in Helsinki the opinions that have been presented here. We are all aware of everything that has been written here about these questions and, therefore, we do not need to waste time for reviews of different opinions. We must also notice that the people of the 20th century are in an enormous hurry towards a still higher living standard. Therefore we must be much more brief than in our discussions in the 17th century. Otherwise these people, pressed by a highly developed bureaucracy, perhaps won't find time to read the report that Mr. Spiritualist is going to publish about our visit.

However, I would like to ask you, my friend Salviati, to briefly describe the situation. What, indeed, is the question in this Finnish discussion?

Salviati: If I must be very brief, I would like to state that the different opinions concern the conception of reality and the question of the eventual limits of the empirical method of research, which was the subject of our enthusiastic discussions in the 17th century: Is this method able to solve all questions concerning reality? You know that science has developed enormously during these 350 years, but especially the investigation of the so-called atomic phenomena which is now so popular has raised the question of whether the empirical method is at all able to reach the properties and features of reality itself. Some physicists have asked whether atoms and elementary particles are real at all or if they are merely theoretical concepts, while essential parts of reality remain for ever beyond our reach. They have claimed that the theoretical models of atomic physics do not approach reality even in an asymptotic sense because reality has *irrational* aspects: It contains features which cannot be reached by the rational methods of science.

Simplicio: As we know, Finnish philosophers have almost unanimously rejected those claims of the 'irrationality' of reality. They have good reasons for this. In our previous discussions I was somewhat critical of the empirical methods recommended by our Academician. The subsequent development has, however, without doubt shown that empirical methods are very effective indeed. But even now, as in the 17th century, I find it necessary to emphasize the criticism presented by philosophers. It reflects even now the general patterns of thought of the establishment. It is true, the clerical establishment has since been replaced by an atheistic establishment. This I find a little disturbing – but we should still pay attention to what the philosophers have to say.

If somebody claims that the scientific research method has limitations because of some kind of 'irrational element' in reality, it is reasonable to require that he present arguments for the real existence of such 'irrational matters'. One has to present a philosophically tenable argument which shows that such matters are not just imagined. As we know, such an argument has not been presented and, therefore, it seems to be just a question of a desire to find a place for God in the scientific picture of the world, and such a desire is alien to science.

Even this endeavor does not seem to be successful. Theologians have, with good reason, pointed out that fortunately God is not irrational. They also say that in this way God is made very small. If one places him into the openings which can now be found in the scientific picture of the world, His state will become more and more uncomfortable in the future as science progresses. Most theologians, therefore, seem to reject the pondering of the relation between knowledge and belief based on such considerations – and this I find quite justifiable.

Sagredo: Maybe it is best that we discuss one question at a time. Let us ignore theological questions today and concentrate on the question of the claimed limitations of the scientific method and the irrationality of reality, *purely from the point of view of the philosophy of science.* What do you, our friend Salviati, who so well understand the nature of empirical research, what do you think of the arguments which have been presented in this respect on the basis of atomic physics?

Salviati: This question can be approached in different ways. The starting point of Niels Bohr was the idea of *complementarity* characteristic of quantum theory. In the delineation of the atomic world, one needs complementary ways of description. As these complementary ways have mutually contradicting properties, Bohr arrived at the idea that we meet some 'irrationality' in the atomic theory. This he states in his famous Como lecture (1927).

Later, however, Bohr changed a little the expressions which he used in describing the matter – probably because of the strong criticism presented by philosophers. He did not use the term 'irrationality' any more, nor did he emphasize the limits of the scientific method but pointed out, on the contrary, that science must be *objective*: one should not mix science with any kind of mysticism which – as he said – is completely alien to the basic nature of science

Wolfgang Pauli, who had a stronger influence on the development of the atomic theory than one might gather from his publications, described these matters in a different way. According to him, it is characteristic of quantum mechanics that one has to reject determinism, i.e, the idea of absolute causality, in the atomic theory. One is forced to use probabilistic laws and, therefore, one must speak of *statistical causality* which, in the atomic theory, replaces absolute causality.

This means that *it is possible to govern statistical mean values only with the aid of a rational theory*, while individual events always contain 'random' oscillation or scattering which is not predictable by any laws. Thus, causality is converted into a form which predicts only summarily the mean behavior while individual events contain real indeterminism (freedom from all laws). It is as if *choices* on the part of nature would take place in individual phenomena, and these 'choices' cannot be explained in any rational way.

One can say that the scientific method does not reach 'the unique', i.e., individual phenomena. These 'fall through the mesh of the net of science'. In individual phenomena the *irrationality of reality* appears: A rational theory cannot describe in detail that which really takes place. If we admit that reality is not governed by any complete laws but all phenomena are governed by statistical causality, then it seems to be indisputable that certain limits of rational knowledge are met here and that one can speak of the irrationality of reality.

Simplicio: Here you, my friend Salviati, neglect the philosophical remarks which I already mentioned. You see, this line of argument does not prove at all that there is something irrational in reality. There seems to appear some random 'wavering' in individual events which cannot be bound to any laws – that I cannot deny – but this does not justify the claim that there is something really irrational in reality or that this wavering would presuppose some kind of 'metaphysical cause'. It is not right to claim that reality contains something that rational theories would not reach. At first one should give evidence of *something that is* behind this observed 'wavering'. As far as this has not been shown, all talk of an 'irrational element of reality' must be considered unscientific.

I admit that the random dispersion characteristic of individual events 'limits' in a certain sense that which we can scientifically govern, but this does not imply any real limits for the scientific method. It has not been shown that there is something beyond those limits. It seems that claims concerning the 'limits' of the scientific method and the 'irrationality of reality' are associated with a *belief* that reality contains something mystical and supernatural. The scientific critique does not give support to such a belief. At the most, an agnostic attitude is acceptable in science. This means tolerance with respect to religiosity, however, rejecting all unscientific speculations concerning matters that man does not know and cannot know.

Salviati: You speak now of science in a way which I cannot accept. If you require that the existence of the irrational element must be 'proved', you certainly mean a proof in the sense of rational science – something that we have learned to require in science. It is not possible, however, to prove the existence of something 'irrational' by using rational methods! The requirement which you present shows that you have fortified yourself into rationality in such a way that nothing can rock your *belief* in the complete rationality of reality. I would like to call this a religious decision which cannot be shaken by any arguments. I would like to call your belief *scientism*.

The scientific critique does not require and does not justify this kind of attitude. It is quite possible that the whole of reality cannot be caught with the net of science. Pauli states that *unique* events go through the mesh of the net of science. The situation encountered in atomic research gives reason for pondering the possibility that there are events which must be considered real but which scientific methods are not able to describe. In fact, all individual events seem to contain such an 'irrational element'. It is meaningless to require that the existence of this irrationality must be proved by using rational methods, since the irrationality explicitly means something to which rational methods cannot be applied.

By the way, C.G. Jung has, in his article concerning *synchronicity,* presented examples of events which have been important but which cannot be repeated (*Jung & Pauli* 1952); thus, they cannot be investigated by the customary methods of empirical science.

Simplicio: But Salviati, do you believe in all kinds of mystical and supernatural phenomena? If we do not require that what we believe in can be scientifically verified, then the way is open to all manner of medieval fancies. Do you find it possible that there are good and evil spirits, which people must try to conciliate and persuade by offerings and prayers in order to succeed in their endeavors? Western people, at least, have an experience of over 300 years of scientific facts countering such beliefs of dark superstition.

Salviati: We are in obvious danger here of running into emotional controversies which are generally associated with disputes concerning the foundations of belief.

Let us, however, take these matters from the point of view of real scientific criticism and avoid emotional remarks. The simile about the net of science I find very illuminating. If one has never caught small fishes with a certain net, one cannot draw the conclusion that there are no small fishes in the lake. The small fishes pass, of course, through the mesh of the net. In the same way, the requirement of rationality which is characteristic of the scientific method limits the possibilities of scientific research so that it is not applicable to all matters which, however, must be called real. On the basis of the results of atomic physics in the 20th century we have full reason to think that this indeed is the case. The scientific criticism does not justify stopping the debate on such questions at the very outset by stating that the discussion becomes 'unscientific'.

If you, Simplicio, require that the existence of the irrationality has to be proved by rational means, by the same token one can require from you an argument which justifies *why* you only believe in such matters which can be shown to exist by rational methods. Neither of these attitudes can be scientifically motivated: One needs a religious decision. I would say that scientism is a dogmatic attitude which is stifling a discussion that is truly needed just now.

Sagredo: I am inclined for my part to accept your remarks. There is reason to deliberate seriously about the irrationality of reality on the basis of the facts which have come up in atomic research. On the other hand, Simplicio has with full justification made the remark that one then very easily opens the door to the realm of pure superstition. Is it not a real danger? Against this background I can understand that many people reject the idea of irrationality as 'unscientific'.

Simplicio: That was exactly what I meant. If we consider as real something that cannot in an acceptable way – let us say, 'in a rational way' – be shown to be real, why do we not then believe in bugbears and fairies? Where is the borderline between the 'real' irrational element and an illusion?

Salviati: Dangerous possibilities are opened here, it is true. But there are dangers in this world, and one cannot eliminate them by just avoiding speaking of them. For example, the requirement of the objectivity of science is often emphasized very strongly. It is as if one wishes to ward off the dangers which might follow if science had subjective aspects.

It is true that the idea of the irrationality of reality implies that we must also consider seriously statements that are based on intuitive and subjective views. This is *the other way* which alone, according to the teachings of mystics, can yield a real comprehension of reality. We know, however, that this way is subjective and leads to controversies. The different views of prophets have caused much evil and suffering because of these controversies. Many people think that objective science is the only way to avoid them.

Critical deliberation has led me, however, to the conclusion that we cannot exclude the way of intuition and mysticism. If we accept – at least as a serious possibility – the idea that rational means do not reach the whole of reality, then we must accept the way of intuition along with the rational way of science if we wish to shape a general view of reality. This is the way of art and of religion – a way of getting engrossed in the inner world. The facts of atomic physics also point out that the rational way is not sufficient for comprehending the nature of reality. Stating that subjectivity is dangerous is not a sufficient argument for the rejection of the other way. The world *is* dangerous, and if we do not wish to see the dangers, we can be driven into yet greater dangers than by meeting them with open eyes.

Simplicio: Now I begin to understand that your way of thought is not based on mysticism but on critical deliberation. It is obvious, I agree, that the strong development of science has brought visible dangers which perhaps are related to the fact that science, one-sidedly, emphasizes rationality. Questions concerning morals and values are largely left out of consideration. Such questions are now becoming important because of the obvious dangers which the development of science implies. Let us only think of the moral problems created by the development of biology, medicine, and data processing – not to mention nuclear physics. Did you mean such questions when you remarked that dangers are not eliminated by leaving them out of consideration?

Salviati: That's exactly what I meant. Pauli emphasizes in his letters that in modern physics matter is treated in a one-sided way, since consciousness is considered as something quite outside the world of matter. The real situation is such that we cannot know anything about the world of matter without the 'interaction' between matter and consciousness in each observation. This fact should be taken into consideration in physics. The conception of reality in modern natural science considers consciousness as 'detached', as something unessential. The result is a one-sided and distorted picture of the world. It is to be feared that technologies which are based on such a conception of reality will wipe out the 'unessential' spirit (consciousness) from the terrestrial globe!

Sagredo: Thus, we meet here the problem of the role of the spirit in the world. Present natural science speaks, indeed, only of matter and related phenomena. The *world of spirit* is something that does not interest natural scientists, and it is quite usual that people have begun to consider the very concept of 'spirit' quite unnecessary. A materialistic attitude is generally associated with science, and this results in scientism.

Is it not so that if we begin to discuss the irrationality of reality we must, simultaneously, also speak of the role of spirit in the world and of the relation between matter and spirit? I mean that we must consider seriously the possibility that one has perhaps to consider spirit to be an element of reality which has an equal importance – and perhaps has an even more important role in reality – than matter. Then one could not consider spirit just as a product of the world of matter – as one is accustomed to think according to the materialistic philosophy.

At this moment, there was a knock on the door and the Head Waiter entered the room.

Head Waiter: Esteemed Excellencies! I have reserved this room for you at the request of Mr. Spiritualist. Unfortunately, I cannot see you, but at my request some electronic experts have wirelessly coupled a listening device to the apparatus of Mr. Spiritualist, and so I have been able to listen to your interesting conversation. However, now you are entering a realm such as that which I feared beforehand. I must make it clear to you, dear Excellencies, that in Finland not all philosophies are quite appropriate. Only in churches and in other meeting places of religious communities is it appropriate to speak of the matters of *spirit*.

I hope that your Excellencies understand these principles which Finnish self-criticism dictates. Referring to these principles, I must ask you that you continue your discussion in a meeting place which is better suited to a discussion concerning matters of spirit.

I hope that you don't misunderstand me. I wish to assure you that I shall willingly reserve a room for you even in the future, but only on the condition that you limit your discussion to purely scientific questions – or to conversations concerning such clear and definite questions as weather or politics, etc. If you, however, explicitly wish to discuss questions which mix science and religion, then I – being responsible for the reputation of this restaurant – must ask you to find another meeting place. By the way, I have the impression that it will not be easy to find a suitable place for such discussions in Finland. I know that at least within the Lutheran church – to which I belong – it has not been deemed proper that such questions be discussed in its meeting places. Let us have science as science and religion as religion. It is not advisable to mix them.

I hope that you understand that I must consider these viewpoints. In fact, I must say that I do not quite understand such matters, but in this way I have understood the unexpressed principles by which I must abide.

Sagredo: This was not entirely unexpected, of course. I think, we all know the term 'finlandization'. In fact, we can finnish this meeting for today now. Tomorrow we can meet again in this same room at the same time if we take into account the fact of finlandization. Thank you, dear Head Waiter, and please reserve this room for us for tomorrow.

References

Arzt et al. 1992 *Unus Mundus: Kosmos und Sympathie; Beiträge zum Gedanken der Einheit von Mensch und Kosmos,* eds. Thomas Arzt, Maria Hippius-Gräfin Dürckheim & Roland Dollinger. Peter Lang, Frankfurt a.M.

Arzt et al. 1996 *Philosophia Naturalis: Beiträge zu einer zeitgemässen Naturphilosophie,* eds. Thomas Arzt, Roland Dollinger & Maria Hippius-Gräfin Dürckheim. Königshausen & Neumann, Würzburg.

Atmanspacher et al. 1995 *Der Pauli-Jung-Dialog und seine Bedeutung für die moderne Wissenschaft,* eds. H. Atmanspacher, H. Primas & E. Wertenschlag-Birkhäuser. Springer, Berlin.

Bateson 1987 Gregory Bateson & Mary Catherine Bateson: *Angels Fear: Towards an Epistemology of the Sacred.* Macmillan, New York.

Bohr 1934 Niels Bohr: *Atomic Physics and Human Knowledge.* Cambridge Univ. Press, Cambridge.

Bohr 1955 Niels Bohr: "The Unity of Knowledge" in *The Unity of Knowledge,* ed. Lewis Leary. Doubleday and Co, New York; reprinted as article 6 in *Bohr* 1985.

Bohr 1963 Niels Bohr: *Essays 1958/1962 on Atomic Physics and Human Knowledge,* ed. Aage Bohr. Wiley, New York.

Bohr 1985 Niels Bohr: *Atomphysik und menschliche Erkenntnis,* ed. Karl v. Meyenn. Vieweg, Braunschweig. [Essays 1930–1961.]

Born 1961 Max Born: "Bemerkungen zur statistischen Deutung der Quantenmechanik" in *Werner Heisenberg und die Physik unserer Zeit,* ed. F. Bopp. Vieweg & Sohn, Braunschweig.

Born 1983 Max Born: *Physik im Wandel meiner Zeit,* eds. Roman U. Sexl & Karl v. Meyenn. Vieweg, Braunschweig.

Burtt 1980 E.A. Burtt: *The Metaphysical Foundations of Modern Physics.* Routledge & Kegan Paul, London (1st ed. 1924).

d'Espagnat 1980 Bernard d'Espagnat: *A la recherche du reel. Le regard d'un physicien.* Bordas, Paris (1st ed. 1979).

d'Espagnat 1983a Bernard d'Espagnat: *In Search of Reality.* Springer, New York. (1st French edition 1979.)

d'Espagnat 1983b Bernard d'Espagnat: *Auf der Suche nach dem Wirklichen.* Springer, Berlin.

d'Espagnat 1993 Bernard d'Espagnat: "Open Realism" in *Symposia* 1992. In Finnish: "Avoin realismi". *Arkhimedes* 2/1993.

d'Espagnat 1994 Bernard d'Espagnat: *Veiled Reality*. Addison-Wesley, Reading, MA.

Eccles 1994 Sir John Eccles: *How the Self Controls Its Brain*. Springer, Berlin.

ECST 1990 *Science and Religion. One World – Changing Perspectives on Reality* (Second European Conference on Science and Religion, 1988), eds. Jan Fennema & Iain Paul. Kluwer, Dordrecht.

ECST 1994 *Studies in Science and Theology*, Vol. 2 (Fourth European Conference on Science and Theology, Vatican City, 1992), eds. George V. Coyne, S.J., Karl Schmitz–Moorman, and Christoph Wassermann. Labor et Fides, S.A., Geneva, 1994.

Einstein 1951 *Albert Einstein: Philosopher – Scientist*, ed. P.A. Schilpp. Tudor, New York (1st ed. 1949).

Enz 1995 Charles P. Enz: "Observability and Realism in Modern Experiments with Correlated Quantum Systems". Session of the *International Academy of Philosophy of Sciences* on: "Observability, Unobservability and their Impact on the Issue of Scientific Realism", Parma, Italy, May 24–28, 1995.

Fierz 1954 Markus Fierz: „Über den Ursprung und die Bedeutung der Lehre Isaac Newtons vom absoluten Raum". *Gesnerus*, Sauerländer Aarau, 11, pp. 62–120.

Fine 1986 Arthur Fine: *The Shaky Game: Einstein Realism and the Quantum Theory*. Univ. of Chicago Press, Chicago.

Florenski, Pavel See *Silberer* 1984.

Folse 1985 Henry F. Folse: *The Philosophy of Niels Bohr: The Framework of Complementarity*. North-Holland, Amsterdam.

Ghose et al. 1992a P. Ghose, D. Horne & G. Agarwal: „An Experiment to Throw More Light on Light", *Phys. Lett.* A, 153.

Ghose et al. 1992b P. Ghose, D. Horne & G. Agarwal: "An 'experiment to throw more light on light': implications", *Phys. Lett.* A, 168.

Guitton 1991 Jean Guitton, Grichka Bogdanov & Igor Bogdanov: *Dieu et la science. Verse le métaréalisme*. Grasset, Paris.

Heisenberg 1955 Werner Heisenberg: *Das Weltbild der heutigen Physik*. Rowohlt, Hamburg.

Heisenberg 1959 Werner Heisenberg: *Physik und Philosophie*. Hirzel, Stuttgart. In English: *Physics and Philosophy*. Allen & Unwin, London, 1959.

Hämäläinen et al. 1994 *Keskustelua tieteen rajoista*, eds. Rauno Hämäläinen, K.V.Laurikainen, Jussi Rastas & Karri Sunnarborg. SEFT Report series: HU-SEFT I 1994-04.

Jung 1961 C.G. Jung: *Memories, Dreams, Reflections*, rev.ed. Aniela Jaffé. Collins, Fount Papers.

Jung 1981 C.G. Jung: *Synchronicity: An Acausal Connecting Principle*. Routledge & Kegan Paul, London (1st ed. 1972).

Jung & Pauli 1952 C.G. Jung & Wolfgang Pauli: *Naturerklärung und Psyche*. Rascher, Zürich. In English: *The Interpretation of Nature and the Psyche* (1955).

Kaila 1942 Eino Kaila: *Über den physikalischen Realitätbegriff. Zweiter Beitrag zum logischen Empirismus* = Acta Philosophica Fennica 4, 1941. Helsinki.

Kaila 1943 Eino Kaila: *Syvähenkinen elämä. Keskusteluja viimeisistä kysymyksistä* (The Depths of Spiritual Life. Discussions on the Ultimate Questions). Otava, Helsinki.

Kaila 1950 Eino Kaila: *Zur Metatheorie der Quantenmechanik* = Acta Philosophica *Fennica* 5. Helsinki.

Laurikainen 1978 K.V. Laurikainen: *Fysiikka ja usko* (Physics and Belief). WSOY, Porvoo.

Laurikainen 1985a K.V. Laurikainen: *Atomien tuolla puolen.* Kirjapaja, Helsinki.

Laurikainen 1985b K.V. Laurikainen: "Wolfgang Pauli and the Copenhagen Philosophy" in *Symposium* 1985.

Laurikainen 1987 K.V. Laurikainen: *Tieteen giljotiini* (The Guillotine of Science). Otava, Helsinki.

Laurikainen 1988 K.V. Laurikainen: *Beyond the Atom: The Philosophical Thought of Wolfgang Pauli.* Springer, Berlin, Heidelberg.

Laurikainen 1992 K.V. Laurikainen: „Ontological Implications of Complementarity" in *Nature, Cognition and System II,* ed. Marc E. Carvallo. Kluwer, Dordrecht.

Laurikainen 1993 K.V. Laurikainen: „Remarks on Physics and the Psyche" in *Symposia* 1992.

Laurikainen 1994a K.V. Laurikainen: *Kantista kvanttiin. Filosofiaa fyysikon silmin.* Yliopistopaino, Helsinki.

Laurikainen 1994b K.V. Laurikainen: *Atomien viesti.* Yliopistopaino, Helsinki.

Lehti 1987 Raimo Lehti: *Tanssi auringon ympäri.* Pohjoinen, Oulu.

Lehti 1994 Raimo Lehti: „Syrjässä oleva havaitsija välttämättömyytenä, ideaalina ja kummajaisena" in *Hämäläinen et al.* 1994.

MacKinnon 1982 Edward M. MacKinnon: *Scientific Explanations and Atomic Physics.* The University of Chicago Press, Chicago & London.

Martinson 1956 Harry Martinson: *Aniara.* Bonnier, Stockholm.

Myllykoski 1994 Matti Myllykoski: *Jeesuksen viimeiset päivät* (The Last Days of Jesus). Yliopistopaino, Helsinki.

Newton 1704 Isaac Newton: *Opticks.* Samuel Smith & Benjamin Walford, London.

Niiniluoto 1981 Ilkka Niiniluoto: "Tilastollinen kausaliteetti ja tieteellinen maailmankatsomus". *Tiedepolitiikka* 3 /1981.

Niiniluoto 1984 Ilkka Niiniluoto: *Tiede, filosofia ja maailmankatsomus.* Otava, Helsinki.

Niiniluoto 1993 Ilkka Niiniluoto: "Vastakohtien kautta maailmankuvaan". *Yliopisto* 9 /1993.

Otto 1917 Rudolf Otto: *Das Heilige. Über das Irrationale in der Idee des Göttlichen und sein Verhältnis zum Rationalen.* Breslau. English translation *The Idea of the Holy* in 1923.

Otto 1926 Rudolf Otto: *West-östliche Mystik.* Gotha.

Pauli Letter Collection (in short PLC) Collection of Pauli's correspondence in the Archive of CERN (Organisation européenne pour la recherche nucléaire), CH-1211 Geneva 23, Switzerland

Pauli 1952 Wolfgang Pauli: "Der Einfluss archetypischer Vorstellungen auf die Bildung naturwissenschaftlicher Theorien bei Kepler" in *Jung & Pauli* 1952.

Pauli 1954 Wolfgang Pauli: "Naturwissenschaftliche und erkenntnistheoretische Aspekte der Ideen vom Unbewussten". *Dialectica*, Vol. 8, No 4. Editions du Griffon, La Neuveville, Suisse; reprinted as article 17 in *Pauli* 1984.

Pauli 1984 Wolfgang Pauli: *Physik und Erkenntnistheorie*, ed. Karl v. Meyenn. Vieweg, Braunschweig (1st ed. 1961).

Pauli 1994 Wolfgang Pauli: *Writings on Physics and Philosophy*, ed. Charles P. Enz & Karl v. Meyenn. Springer, Berlin. An English translation of *Pauli* 1984, including the article *Pauli* 1952.

Pauli & Jung 1992 Wolfgang Pauli & C.G. Jung: *Ein Briefwechsel 1932-1958*, ed. C.A. Meier. Springer, Berlin.

Plato 1987 Plato: *The Collected Dialogues*, eds. Edith Hamilton & Huntington Cairns. Princeton Univ. Press, Princeton, N.J. (1st ed. 1961).

Platon 5 1982 Platon: *Teokset, V osa*, suom. Marja Itkonen-Kaila, A.M. Anttila & Marianna Tyni. Otava, Helsinki.

Polkinghorne 1988 John Polkinghorne: "The Quantum World" in *Physics, Philosophy and Theology: A Common Quest for Understanding*, eds. Robert John Russell, William R. Stoeger, S.J. & George V. Coyne, S.J. Vatican Observatory.

Polkinghorne 1994 John Polkinghorne: *Science and Christian Belief. Theological Reflections of a Bottom-up Thinker*. SPCK, London.

Popper 1982a Karl R. Popper: *The Open Universe. An argument for Indeterminism*, ed. W.W.Bartley, III. Hutchinson, London.

Popper 1982b Karl R. Popper: *Quantum Theory and the Schism in Physics*, ed. W.W.Bartley, III. Hutchinson, London.

Primas 1990 Hans Primas: "Beyond Baconian Quantum Physics" in *Kohti uutta todellisuuskäsitystä. Juhlakirja professori Laurikaisen 75-vuotispäivänä* (Towards a New Conception of Reality. Anniversary Publication to Professor Laurikainen's 75th Birthday). Yliopistopaino, Helsinki.

Primas 1992 Hans Primas: „Umdenken in der Naturwissenschaft" in *GAIA, Ecological Perspectives in Science, Humanities, and Economics*. Heidelberg.

Primas 1995 Hans Primas: „Über dunkle Aspekte der Naturwissenschaft" in *Atmanspacher et al.* 1995.

Rosenfeld 1979 *Selected Papers of Léon Rosenfeld*, eds. R.S. Cohen & J.J. Stachel. Reidel, Dordrecht.

Selleri 1990 Franco Selleri: *Die Debatte um die Quantentheorie*. Vieweg, Braunschweig (1st ed. 1983).

Shimony 1992a Abner Shimony: "Physical and Philosophical Issues in the Bohr-Einstein Debate" in *Symposia* 1992.

Shimony 1992b   Abner Shimony: "Remarks on the Approaches to Quantum Mechanics" in *Symposia* 1992.

Shimony 1993   Abner Shimony: "Physical and Philosophical Issues in the Bohr-Einstein Debate" and "Remarks on the Approaches to Quantum Mechanics" in *Symposia* 1992.

Silberer 1984   Michael Silberer: *Die Trinitätsidee im Werk von Pavel A. Florenskij. Versuch einer systematischen Darstellung in Begegnung mit Thomas von Aquin.* (Das östliche Christentum, Neue Folge, Band 36.) Augustinus-Verlag, Würzburg.

Stapp 1993   Henry P. Stapp: *Mind, Matter, and Quantum Mechanics.* Springer, Berlin.

Stenholm 1986   Stig Stenholm: *K.V. Laurikainen: Atomien tuolla puolen.* Arkhimedes 1A/1986, pp. A101–A104.

Symposia 1992   *Symposia on the Foundations of Modern Physics 1992: The Copenhagen Interpretation and Wolfgang Pauli,* eds. K.V. Laurikainen & Claus Montonen. World Scientific, Singapore, 1993.

Symposium 1985   *Symposium on the Foundations of Modern Physics: 50 Years of the Einstein–Podolsky–Rosen Gedankenexperiment* (1985), eds. P. Mittelstaedt & P. Lahti. World Scientific, Singapore, 1985.

Symposium 1987   *Symposium on the Foundations of Modern Physics: The Copenhagen Interpretation 60 Years after the Como Lecture,* 6–8 August 1987, Joensuu, Finland. *Discussion Sections,* eds. P. Lahti, K.V. Laurikainen & J. Viiri. Report Series Turku-FTL-L45.

Symposium 1994   *Symposium on the Foundations of Modern Physics: 70 Years of Matter Waves,* eds. K.V. Laurikainen, Claus Montonen & Karri Sunnarborg. Editions Frontières, Paris, 1994.

Työrinoja 1994   Reijo Työrinoja: „Teologia - tiede - teknologia" in *Suomalaisen Teologisen Kirjallisuusseuran vuosikirja 1994,* ed. Petri Järveläinen. Helsinki.

Van Erkelens 1993   Herbert Van Erkelens: "Modern Physics and the Symbols of the Self" in *Symposia* 1992.

Weinberg 1978   Steven Weinberg: *Kolme ensimmäistä minuuttia. Moderni näkemys maailman synnystä.* Tammi, Helsinki (in English 1977).

Weinberg 1993   Steven Weinberg: *Dreams of a final theory.* Vintage, London.

Wheeler 1980   John Archibald Wheeler: „Delayed–Choice Experiments and the Bohr–Einstein Dialog" at the joint meeting of the American Philosophical Society and the Royal Society, London, June 5, 1980.

Wigner 1960   Eugene P. Wigner: "The Unreasonable Effectiveness of Mathematics in the Natural Sciences". *Communications in Pure and Applied Mathematics* 13/1. Reprinted as article 17 in *Wigner* 1967.

Wigner 1962   E.P. Wigner: "Remarks on the Mind–Body Question" in *The Scientist Speculates,* ed. I.J. Good. Heinemann, London, 1961; Basic Books, New York, 1962; reprinted in *Wigner* 1967.

Wigner 1967   Eugene P. Wigner: *Symmetries and Reflections.* Indiana Univ. Press, Bloomington.

Abridged Index

Index concerns Parts I–IV in the text. For frequently appearing title words only the most characteristic entries are given.